PRAISE FOR *THE PRICING MODEL REVOLUTION*

"There are many books on pricing. This is the best read for managers wanting a review of several innovative pricing methods."
>—Philip Kotler, Professor of International Marketing, Kellogg School of Management

"Zatta's book is a must-read for anyone who wants to price successfully in the future: it offers many insights and concrete ideas on a topic of fundamental importance for business profitability."
>—Thomas Ingelfinger, Board Member, Nivea-Beiersdorf

"The book of Danilo Zatta is a unique and visionary travel on the discovery of the 'pricing code.'"
>—Il Sole 24 Ore, leading Italian business newspaper

"No matter what external environment you operate in, what product you offer, what customer need you fulfill: this book written by a true pricing expert will inspire you to find the optimal and flexible pricing model to maximize value for both your business and your customers."
>—Alessandro Piccinini, CEO Nespresso Austria

"Having worked successfully with Danilo Zatta in the past I am really looking forward to his new book and inspiring, innovative ideas on pricing as key strategic lever."
>—Christoph Berens von Rautenfeld, Director Competitiveness Program, Siemens Smart Infrastructure

THE PRICING MODEL REVOLUTION

THE PRICING MODEL REVOLUTION

REVOLUTION

HOW PRICING WILL CHANGE
THE WAY WE SELL AND BUY
ON AND OFFLINE

DANILO ZATTA

WILEY

Registered office
John Wiley & Sons, Inc., 111 River Street, Hoboken, NJ 07030, USA

John Wiley & Sons Ltd, The Atrium, Southern Gate, Chichester, West Sussex, PO19 8SQ, United Kingdom

Editorial Office
John Wiley & Sons Ltd, The Atrium, Southern Gate, Chichester, West Sussex, PO19 8SQ, United Kingdom

For details of our global editorial offices, customer services, and more information about Wiley products visit us at www.wiley.com.

Library of Congress Cataloging-in-Publication Data is Available:

ISBN 9781119900573 (Hardback)
ISBN 9781119901150 (ePub)
ISBN 9781119901167 (ePDF)

Cover Design: Wiley
Cover Image: ©Symbolon/ The Noun Project, Inc.
Author Photo: Courtesy of the Author

To my wife Babette and my children Natalie, Sebastian, and Marilena who charge me with energy every day.

In memory of my mother Annemarie – I will hold you forever in my heart.

CONTENTS

PREFACE

"Let's go invent tomorrow rather than worrying about what happened yesterday."

—Steve Jobs

More and more C-level executives are realizing the key role that professional price management plays. From an operational prerequisite to conduct business, pricing became first in the United States, then also in Europe, Asia, Middle East, and Africa a key priority on top of the CEO agenda. Indeed, it is proven that companies where pricing is a C-level priority outperform their peers in terms of profitability.

The context of technological advances and data science progress, paired with new ecosystems and new marketing frontiers, are disrupting old revenue models, accelerating what we call the Pricing Model Revolution: an innovative way to capture the value delivered by the companies to their clients.

Developing new pricing models often means turning around a situation of decreasing revenues and profits to return on a path of profitable growth. It also helps establishing a competitive advantage, vis-à-vis companies sticking to the fading transactional world and applying old school pricing.

The goal of the book is to illustrate the many new routes to corporate profitability that innovative monetization approaches offer. In Part I, we start with the background and context of the Pricing Model

Revolution. In Part II, ten of those approaches will be detailed, always following a three-step approach: first, a *case history* illustrates real-life applications of the approach presented in the chapter. Second, in the *analysis of context* section a deep dive into the topic is presented. Third, the key learnings of the chapter are highlighted in the *summary*. In Part III we illustrate how to win within the Pricing Model Revolution.

This book intends to serve as a source of inspiration and as a brainstorming platform that provides several real-life case studies, anecdotes as well as corporate monetization and pricing model examples to help the reader find their way to improve their own monetization approach.

—Danilo Zatta
Rome/Munich, May 2022
danilo.zatta@alumni.insead.edu

ACKNOWLEDGMENTS

I feel extremely lucky to work on pricing and monetization topics that are innovative and of strategic relevance. I also feel privileged that I am able to work with companies and investors across all industries and geographies to help them prepare for the future and create strategies that will enable them to profitably grow. This advisory work allows me to learn every day, and a book like this wouldn't have been possible without it.

I would like to acknowledge the many managers who have helped me get to where I am today – all the great people in the companies I have worked with who put their trust in me to help them and in return gave me so much new knowledge and experience. I must also thank everyone who has shared their thinking with me and has allowed me to collect and quote case studies as well as concrete examples of successful innovations on the monetization front. I would like to express my most sincere gratitude to them. I am also lucky to personally know many of the key thinkers, pricing experts, and thought leaders in business and I hope you all know how much I value your inputs and our exchanges.

I would also like to thank monetization passionates, practitioners, CEOs, advisors, and sparring partners for the enriching discussions and deep dives on all aspects of pricing (in order of their contributions): Kilian Fleisch, Philip Kotler, Silvia Cifre-Wibrow, Thorsten Lips, Ineke Wessendorf, Francesco Quartuccio, Gábor Ádám, Haarjeev Kandhari, Benjamin Schwarzer, Kai-Markus

ACKNOWLEDGMENTS

Müller, Benjamin Grether, Mátyás Markovics, Mauro Garofalo, Patricia Hampton, Markus Czauderna, Anna van Keßel, Helmut Ahr, Axel Borcherding, Ralf Gaydoul, Ueli Teuscher, Giovanni Battista Vacchi, Thomas Ingelfinger, Vittorio Bertazzoni, Christoph Berens von Rautenfeld, Alessandro Piccinini, Frank Göller, Alf Neugebauer, Luigi Colavolpe, Dietmar Voggenreiter, Paolo De Angeli, and Simone Dominici.

I would like to thank my editorial and publishing team for all your help and support. Taking any book from idea to publication is a team effort and I really appreciate your input and help – thank you, Annie Knight, Debbie Schindlar, Corissa Hollenbeck and Laura Cooksley for having supported with enthusiasm this editorial project right from the beginning.

My biggest acknowledgment goes to my wife, Babette, and our three children, Natalie, Sebastian, and Marilena, for giving me the inspiration, motivation, and space to do what I love: learning and sharing ideas that will help companies growing and prospering.

PART I
THE PRICING MODEL REVOLUTION

CHAPTER 1
MONETIZATION AS A PRIORITY

"The single most important decision in evaluating a business is pricing power.

If you've got the power to raise prices without losing business to a competitor, you've got a very good business.

And if you have to have a prayer session before raising the price by 10 percent, then you've got a terrible business."

Warren Buffett, President of Berkshire Hathaway

Pricing: The New Source of Competitive Advantage

The most successful companies – those with above-average profits – have discovered the new source of competitive edge: pricing and, along with it, the way they can capture the value they provide to their customers through innovative monetization approaches.

Despite being the main and strongest profit driver, in many companies, pricing still remains at the stage of mere potential. This does not enable full profit to be obtained, whilst in the worst of hypotheses an inadequate pricing model ends up by losing customers and with them revenues and profits.

We set prices: "As we always have done" or again: "Adding our margin to the base cost" – these are typical statements from the old world when sales were purely transactional. "I give you a product x and you give me y dollars" was the mantra then. In a context where demand outweighed supply, where customers' demands were unsophisticated, where competitors were more or less analogous and technology not widespread, this might have been a sustainable approach. Not today. The time has come for a change.

There are, however, also companies that realized the importance of pricing, but either lack a structured approach to optimize their monetization, simply ignore or miss out the many pricing levers that if activated could strongly improve their profits or do not have sufficient top management attention on this key topic. The most successful firms are those that have as their number one priority total comprehension of the value perceived by their customers, combined with innovative approaches to monetization.

One of the first things they are quite clear on, is that price is the main profit driver.

If we take the case of a company with fixed costs amounting to US\$30 million, variable costs coming to \$60, a sales turnover of 1 million units and a price equal to \$100, we find ourselves with a profit of \$10 million. If we now improve every profit driver by 1%, in the equation given by profit equal to price by quantity, or the revenue less the costs both fixed and variable, we have the following result: pricing, compared to all three of the other profit drivers, that is, fixed costs at 3%, quantity at 4% and variable costs at 6%, is the lever that makes the greatest impact: increasing profits by as much as 10% (see Table 1.1).

Firms with above-average profits grasped this mechanism some time ago: they know that pricing is not only the most powerful lever but also works the fastest. Whereas on the side of costs, improvements

Table 1.1: Impact of 1% on all profit drivers

	Starting situation	Improvement of 1%	New profits	Increase in profits
Fixed costs	$30,000,000	$29,700,000	10,300,000	3%
Quantity	1,000,000	$1,010,000	10,400,000	4%
Variable costs	$60	$59.4	10,600,000	6%
Price	100	$101	11,000,000	10%

Source: Adapted from Zatta Danilo et al. (2013), Price Management, Franco Angeli, p. 15.

of even a mere 1% can require large investments and take a long time (e.g. for moving production plants to countries with low production costs, etc.), a 1% improvement in pricing can occur instantly and at zero cost (e.g. when digital pricing labels are changed on the retail shelves in just a few seconds and at zero cost).

Once the power of pricing is understood, companies ask themselves which levers to activate to improve their monetization skills. The answer to this question is that there is not a single pricing lever, but several levers that can potentially be activated, as indicated in the Pricing Framework represented in Figure 1.1.[1] These levers can be assigned to four categories.

The first is related to *price strategy* and contains several facets, like the revenue model, positioning, and differentiation. Moreover, the direction set by the company in relation to its monetization priorities is part of this first category: one of the issues addressed is the question whether the company would be prepared or reluctant to sacrifice market share to increase its profits. In the automotive industry, until a few years ago the answer to this question was a clear no: volume and market share ruled. Today, the view on this has changed quite drastically.

Figure 1.1: The Pricing Framework: from price strategy to price steering
Source: Courtesy of Horváth

The second category is about *price setting*. Price logic, portfolio pricing, and product and service pricing are key facets here. If we take price logic as an example we will find several possible approaches – from cost plus pricing to competitive pricing or value pricing – based on the pricing maturity of a company, as indicated in Figure 1.4.

Once the price strategy has been defined and prices are set, in the sales process we see prices moving from the initial list price to the final transactional price. This is the essence of the third category – called *price implementation* – with, for example, terms and conditions provided by the company to its resellers and distributions partners, execution and negotiation of prices. There are also companies selling directly or not selling via price lists, for example, operating in project business with very customized products or services – all these cases are contemplated in this third category as well.

Finally, companies need to monitor and ensure that the target profitability is reached at the end of the year. To ensure this, *price steering* is needed, as indicated in the last category, where price controlling, price analytics, and price reporting come into play.

To ensure that pricing becomes an integral part of the company and is properly embedded there is a supporting layer, called *pricing enablers*. A clearly structured pricing organization, defined pricing processes (e.g. related to yearly price reviews and increases), pricing IT systems, and pricing skills are all relevant enablers.

To show how many and diverse pricing levers companies can activate you can review the elements displayed as examples in Figure 1.2,[2] where their typical profit impact is shown. These elements can vary from industry to industry and in terms of single elements and profit impact. However, the same key learning holds true for all industries: to become more profitable there is not only a single element to be activated on the revenue side. Several pricing levers can be optimized and the sum of the impact of each of them generates a substantial profit improvement.

Triggers of the Pricing Model Revolution

In the last years we observed companies changing their monetization approaches: the most profitable companies were capable of assessing where value was coming from in the eyes of their clients and adapt their monetization approaches accordingly, creating a sustainable competitive advantage.

The period of the 2020–2022 pandemic gave a further boost to change and digitization, opening up to pricing and new models of revenue – barriers that previously seemed impossible to force open.

Starting from the basis of this ground-breaking change in the way companies monetize the value they provide to their customers, there are particular elements that we have grouped into four groups: these are the accelerators, or *triggers* of the Pricing Model Revolution (Figure 1.3). They are changing and will increasingly change the way firms extract and will continue to extract value from the market.

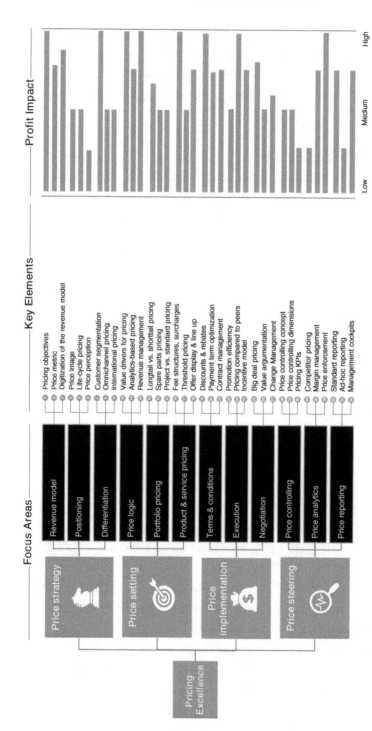

Figure 1.2: Pricing Framework: key elements and profit impact
Source: Courtesy of Horváth

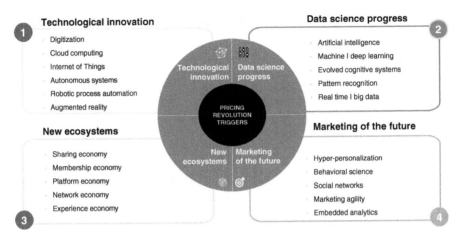

Figure 1.3: The four triggers of the Pricing Model Revolution

The *technological innovation* of the last few years is the first trigger: it has laid the bases for raising pricing to a new level: digitization, cloud computing, the Internet of Things (IoT), autonomous systems, robotic process automation (RPA), or augmented reality. New cloud applications or digital pricing are often the prerequisites for holistic, data-oriented price management.

Data science progress is the second trigger: there are new quantities and a quality of data that create completely new potentials for setting prices. Suffice it to think of the huge quantities of big data available and how this can provide real-time elasticity regarding individual products or optimal discounts thanks to artificial intelligence. What data science can generate today in terms of knowledge in the field of pricing seemed like science fiction only a few years ago.

Today this also happens within *new ecosystems*, the third trigger, centered on the sharing or repeated use of products but without possessing them: this type of ecosystem requires new pricing models that did not exist in the old world of transactional pricing.

Marketing of the future, or Marketing 5.0, concludes the series of the four triggers: hyper-personalization, timidly initiated with the introduction of systems of revenue management in the service sector, assumes new dimensions thanks to the happy combination of technological innovation and progress in data science. The same is true for inspirations coming from the behavioral sciences and agility in marketing.

These four triggers are the basis of the Pricing Model Revolution.

The Pricing Model Revolution

The transactional model based on the possession of a product is a thing of the past – an inferior pricing model in many cases. New and more innovative pricing models focusing instead on the monetization of use or the result produced by the product have proven clearly superior. Their introduction has allowed companies undergoing crisis to reinvent monetization, doing away with resistance to purchasing and welcoming the customer's readiness to pay. How then is the management of this most important profit driver changing?

In Figure 1.4[3] we see how pricing has evolved. Companies using basic pricing are the least profitable. Here we find the absence of any consistent pricing logic: the same price is maintained for a long period. With cost plus pricing, the price setting is purely based on internal reasoning and calculations: a target margin is added to the cost, which equals to the requested price. It is easy to calculated, if the cost structure is solid, but again limited to an internal perspective that ignores competitors and customers. Although competitor pricing does look further, it ignores the value "perceived" by its customers. All this is taken into consideration in value pricing, which is the most complete and promising approach of all those described up to now.

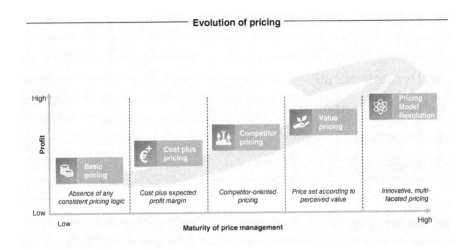

Figure 1.4: Evolution of pricing: from basic pricing to *Pricing Model Revolution*

Source: Courtesy of Horváth

When walking along the maturity stages, companies on average increase their return on sales between 2% and 8% based on their pricing maturity – which typically is a substantial profit improvement.

It is with the Pricing Model Revolution that we attain the Olympus of Pricing.

It represents the evolution of pricing based on value that grows and becomes more sophisticated thanks to approaches to monetization which, in an innovative way, create a firm competitive edge.

In this book we have identified 10 elements that denote innovative monetization approaches that in many cases represent key pillars of a company's competitive advantage:

1. Pay-per-use
2. Subscription

3. Outcome-based pricing

4. Psychological pricing

5. Dynamic pricing

6. Pricing based on artificial intelligence

7. Freemium

8. Sympathetic pricing

9. Participative pricing

10. Neuropricing

These new approaches to monetization are not limited to certain industries or geographical areas: they are all-pervasive and will continue to evolve, increasingly changing the rules of the trading game. They create a transparency in terms of customer needs, product use, and readiness to pay, that has never been available previously. Products are transformed into services. Their value is translated into measurable units of performance.

The Pricing Model Revolution establishes a strategic imperative: a new way of pricing is here and it is the new and unstoppable source of competitive advantage.

Summary

Of all profit drivers – that is, volume, price, and cost – price is not only the strongest one but it is also the one that can be realized faster and more efficiently than all the others.

To increase profitability through pricing companies can activate multiple levers. The Pricing Framework groups the key levers into the following categories: price strategy, price setting, price implementation, and price steering.

In addition, price enablers help embedding pricing properly into an organization.

There is not a single lever but a whole array of pricing levers. The sum of their effects usually adds up to significant profit improvement: on average between 2% and 8% in terms of higher return on sales.

The most profitable companies started innovating their monetization approaches, creating a sustainable competitive advantage leading to a Pricing Model Revolution.

There are four accelerators, or *triggers*, of the Pricing Model Revolution: technological innovation, data science progress, new ecosystems, and marketing of the future.

Ten elements that denote innovative monetization approaches, that in many cases represent key pillars of a company's competitive advantage, are: pay-per-use, subscription, outcome-based pricing, psychological pricing, dynamic pricing, pricing based on artificial intelligence, freemium, sympathetic pricing, participative pricing, and neuropricing.

PART II
NEW APPROACHES TO MONETIZATION

CHAPTER 2
PAY-PER-USE / PER-WASH / PER-MILE /. . .

". . .only try to realize the truth . . .
Then you'll see that it is not the spoon that bends, it is only
yourself."

From the film, *The Matrix*

Case History

External, day.

A soldier is leaning against the window of a laundromat.

A young man enters, takes off his shades. As a child appears from behind a giant washing machine, the guy takes off his T-shirt and jeans, stripping down to his boxer shorts. Appreciative glances from the girls, an elderly lady drops her jaw, as the young man sits down and starts reading the newspaper next to a middle-aged man stoically seated on a chair – the triumph of adolescence over daily routine.

This is the Levi's ad of the 1980s, with Creedence Clearwater Revival's *Heard it Through the Grapevine* playing in the background, starring a young Nick Kamen, the sadly missed musician who made such an impression on Madonna, along with half the rest of the world, that she produced his first record.

The vehicles of an enterprise make their effect thanks to the mood it manages to create. In the same way, it is enough to change the order of the factors to see how that *short story* (the original ad lasted a little over thirty seconds) can be applied to the dynamics of pricing. In particular, how important it is for the sake of change, not necessarily to be left standing in your underwear, but to overturn the rules of what already exists, for example by "showing" that even something dead easy like washing clothes, can attract people, if it's placed in a different context. It's up to us to decide whether we want to deal with the change by interpreting the young man, the kid, the pin-up girls or the guy sitting on the chair.

Now let's imagine being the CEO of a company producing upmarket domestic electrical goods. The company is a global leader in a niche market that enjoys *premium* prices: in fact it typically sells to big business-to-business (B2B) customers, such as hotel and restaurant chains that can afford high-priced products.

At the start, given the product's strategic positioning, their business grows visibly. And new markets are sought: small pubs and restaurants are identified as a new segment that might increase revenue. However, premium prices are already "saturated". There is no way higher prices can be demanded.

This is a conceptual limit, according to the *pay-per-use* view. A little like the kid in the film *The Matrix* who says to the main character, Neo: "Do not try and bend the spoon. That's impossible. Instead . . . only try to realize the truth. There is no spoon. Then you'll see that it is not the spoon that bends, it is only yourself." It's the same with pay-per-use. It's impossible to bend the market. Instead, it's better to bend the policies for approaching the market. Then you'll notice a simple fact: the present pricing model is an obstacle to growth. This is why we must decide to break with the logic of traditional pricing – based on ownership – and find an innovative way for winning new customers.

This, for example, is what was done by the German company Winterhalter, world market leader in the sector of high-quality, commercial dishwashers, when they introduced pay-per-wash.

As Ralph Winterhalter, CEO of Winterhalter, put it, with a new approach to pricing: "You only pay for the dishwasher when you use it[1]: this is particularly important for all those companies that depend on seasonal work, such as beer gardens, mountain resorts, beach bars, where dishwashers are not in use all year round. So "Why invest money in owning a dishwasher you don't even use for half the year?", Ralph Winterhalter asks himself and all potential customers.

And so the company launched the *Next Level Solutions* project, which exploits the latest digital technology in order to bring commercial dishwashing to their new segment of the catering business, which traditionally would not be able to afford such a product.

This produced a dual effect by appealing to exclusivity, too. The customer not only gains a feeling of satisfaction from having paid less, but the process also becomes objectively more efficient and sustainable, whilst from a psychological point of view the message is conveyed that things have been upgraded to "Championship level".

Winterhalter's new pricing model had the objective of offering *premium* dishwashing, independently of the customer's budget, that is, with the advantage for the latter that no initial investment was needed and a zero-risk operation was ensured.

> *Use-based invoicing is calculated by washing cycles.*
> *Detersives and other products are all included.*
> *Post-sales services, such as maintenance and repairs are*
> *also factored in.*

Moreover, the contract with Winterhalter can be terminated at any time, with no obligations, thus offering maximum flexibility.

The Winterhalter case shows just what pay-per-use means: customers can use the goods they need when they need them, without the burden of ownership and without having to pay for periods when these precious assets are inactive. Moreover, they can adapt the mode of use dynamically, scaling it up or down, to adapt to circumstances and unforeseeable future needs, such as preferences in demand, financial position, environmental conditions, and other considerations, according to the ecosystem they operate in.

Without the need for ownership of the asset, customers can devote more time and resources to the efficient use of products (rather than to their installation, maintenance, and updating); furthermore, they can also start using the product immediately, since nowadays distribution channels are becoming quicker and quicker and, in many cases, digital.

Analysis of Context

"Aligning price with use" is the essence of pay-per-use.

The strength of this approach lies in its ability to break down barriers to purchase and expand market potential, orienting companies and allowing them to do business by making innovations to existing models.

Once set up, pay-per-use determines a clear vision of how customers use their products or services. This, for example, generates a deeper understanding of how they can provide even better value, improving their offer according to a broader customer base and creating a platform for growth.

All this allows for more accurate forecasts, value extraction and in some cases modification of product development to better satisfy demands.

Even though the pricing concept based on use is not new and many companies in different sectors adopt it (see Table 2.1), up until a short while ago the costs of the sensors and technology necessary for setting up pay-per-use for smaller or more dynamic increments were prohibitive.

With the increase in digitization, big data, and artificial intelligence, the possibility of "winning" customers on an on-demand basis has become feasible.

Technological progress (high-speed Internet), a drop in the price of microchips and the expansion of cloud-computing capacity make the offer economically sustainable, as well as monitoring and invoicing both for business-to-customer (B2C) and B2B businesses.

Table 2.1: Selected companies that have introduced pay-per-use

Company	Product	Offer	Pricing model
Winterhalter	Domestic electrical appliances	Next Level Solution	Pay-per-wash
Rolls Royce	Aeronautic engines	Total Care	Pay-per-hour-flown
Atlas Copco	Compressed air	Air Plan	Pay-per-m^3
Zipcar	Mobility	Car Sharing	Pay-per-hour
Amazon Web Service	IT services	Cloud Computing	Pay-per-GB
Michelin	Tires	Michelin Effitires	Pay-per-mile
Samoa Air	Air transport	Intera offerta	Pay-per-kilo

All these developments have favored a key factor in the spread of pay-per-use models, which is the ability to capture latent demand by reducing the initial cost associated with physical assets for customers with a low rate of use.

This triggers market expansion: new segments of customers who, using a traditional pricing model, would not have had the chance or intention to buy the product, can now afford to use it.

And there is more.

This phenomenon, combined with the product's shorter life span, generates a further demand for flexible and scalable options at low risk compared to the "traditional" concept of ownership.

The pricing (alignment) of the product and its use can radically overturn and transform the industrial structures and strategies of a company's go-to-market: both actual and potential customers reconsider how, where, and when a product is used. As soon as the products are available on request, with small increments and without having to sustain a large initial investment, more potential buyers will be available.

To unleash the market's latent demand, a growing number of pay-by-use applications have been adopted by companies all over the world: *pay-per-wash, pay-per-ride, pay-per-cleaned-square-meter, pay-per-exercise, pay-per-processing capacity* or *pay-per-mile* are just the beginning of what on-demand pricing can do.

Let's take a closer look at some applications of this type of motivation.

Paying by "Cleaned" Square Meters

Facility management operators in general, and more specifically cleaning companies, traditionally operate at fixed prices.

They may, for example, offer cleaning services at a fixed price per structure: all the spaces regularly cleaned over an established period of time. That's it.

But this sector, too, has experienced the advent of new pricing models: for example, the system that foresees paying per square meter cleaned.

Facility management companies are thus changing their revenue models. New technologies are revolutionizing facility management, making processes more efficient. Why should an unused office be cleaned? Sensors can tell operators which offices have been used and which haven't. A complete set of equipment and detersives can be directly factored into the price per square meter cleaned, which also makes the life of the facility-management staff easier. In this way, only those offices that have been used are cleaned and payment is made per square meter cleaned. This speeds up and optimizes cleaning.

Kärcher, a German, family-run company, has become world leader in cleaning technology, with 100 subsidiaries in 60 countries, and also introduced an innovative pricing model. They call this trend "cleaning on demand."[2]

Paying According to Individual Exercise

Every time we enroll at a gym, we optimistically think the same thing, "This time I'll work out every day." And as usual, we overestimate ourselves. Then, as usual, the week before we start we go on a spending spree: the whole Olympic athlete kit (!), including, of course, those white trousers we fancied and that professional T-shirt and even the elasticated supports to avoid strained muscles. Then something comes up, like a meeting, a supper date: "OK, just for today I'll give it a miss." Next time there's an apéritif: "Shucks!" And bang goes the catch-up session, too.

No extra lengths and goodbye to your good intentions with trim belly and perfect abs included in the price: so what you're looking for is "pay-per-exercise" or "gym-as-you-go".[3] This form of payment, based on use rather than on a monthly subscription, links your enrollment to the actual use you make of it.[4]

Technological progress has made this pricing model possible. Here's how it works.

Near-field communication, a combination of communication protocols for two electronic devices – for example, a smartphone and a piece of gym equipment – allows subscribers to check in directly on the equipment they use for the workout.

The athlete is then charged for the time they use the equipment.

There is no charge for any subscription or membership and those enrolled in the scheme can start or stop whenever they wish.

Today an increasing number of machines are fitted with some form of built-in, short-range communication, to allow those who enroll in a gym, for example, to follow their own training schedule, and this means the time is ripe for the application of this type of pricing.

On the one hand, the price lists provide an option for those who rarely work out, so to waste less money, and on the other hand, gyms manage to attract a different sort of easier-going customer less sensitive to a relatively high price.

In this way, gyms can also orientate the demand for certain machines that can be made constantly available through *surge pricing*, which is obtained when people request immediate access because they don't have the time to wait around and are therefore willing to pay for the use of that specific piece of machinery.

This pricing model opens up new horizons of demand management for operators in gyms: they can reduce prices at off-peak times to make attendance more uniform throughout the week and avoid over-crowding. Moreover, in this way gyms can get an immediate idea of which machines are the most popular and how intense their usage is. This allows them to adapt their range of equipment, for example by purchasing more of the most popular machines, keeping up their maintenance and even running targeted marketing campaigns on the basis of the work load.

Some gyms might be concerned about cannibalization, that is, losing income when members choose not to pay for subscriptions they fail to take advantage of, in favor of *à la carte* prices. But where so many gyms compete with one another in the big cities, alternatives to monthly subscriptions can be a powerful means of distinguishing a business.

Paying for the Skill of Elaboration

By aligning the pricing of products and/or services according to their use, several of a customer's requirements are satisfied, whether it's a matter of flexibility or perhaps high growth, which makes it necessary to modify company policies according to the ups and downs of the market (as happened during the long period of the Covid-19 pandemic when gyms and swimming pools together with many other businesses, remained closed for over a year) or any other unforeseen factors.

Managing to reduce the negative impact of this *alea*, so greatly variable in the area of economics, at least since the beginning of the financial crisis (which began in 2007 with the US subprimes), would be impossible in terms of cost for clients who had to face individual purchases of the infrastructures they needed to satisfy their requirements.

The European Space Agency's Gaia Programme is a good example.

This initiative came into being with the ambitious objective of creating the biggest and most accurate 3D map of the galaxy.

The prerequisite for this praiseworthy venture was the elaboration of satellite observations of over one billion stars. The investment necessary for creating internal capacity sufficient for this sort of data elaboration was estimated to be over US$1.8 million. Nonetheless, the agency only required this particular capacity for two weeks every six months.

To tackle this huge set of data the European Space Agency chose to pay Amazon Web Services for the elaboration of six years' work and the observation of something like a billion stars, which resulted in spending less than half the sum allocated.

By means of a pay-per-use scheme, the products planned for purchase because they are essential for the infrastructures are "re-oriented" as services.

The same goes for Amazon Web Services, known as AWS, which offers cloud-computing, on-demand services to individuals, companies, and public institutions, and charges according to the gigabytes transferred.

Paying by the Mile

The alignment of pricing with usage generates benefits for those clients who use a product infrequently or unpredictably.

This fact has been acknowledged by insurance companies. Thanks to technological progress, the cost of developing small wireless devices capable of monitoring kilometers travelled, by connecting them to the diagnostic port of an automobile has become negligible. And so, companies like MetroMile offer their customers car insurance on a per-kilometer basis, making it an economic proposition for occasional drivers to be able to enjoy full insurance strictly for the time they're using the vehicle: the average saving according to MetroMile is 47%.[5]

It should be noted that in general all the models of on-demand pricing allow for increasingly informed choices, thanks to which clients can test a product and get an idea of its use, what is more without a high initial outlay.

Another case of payment "according to mileage" is offered by Michelin,[6] a leading producer of tires. After developing innovative tires for commercial vehicles, claiming to last 25% longer compared to those of its competitors, the company also realized that it could not apply a 25% increase to its price list and the sales division reacted by advising not to settle for the percentage-price correlation.

Michelin therefore decided to review the company's monetization model: why not link the performance of the tire to its price? The shift from a price-per-tire model to one of price per kilometer involved a classic pay-per-use formula enabled by GPS technology directly connected to the vehicle, in which the entire added value of innovation could be monetized. The longer the tire lasted, the greater Michelin's revenue would be in this case.

Over time, Michelin pushed further ahead: today they provide complete solutions to companies in all sectors, with per-kilometer models for motor vehicles, number of landings for airlines and tons transported in the sector of mining transportation.[7]

Michelin has thus moved from being a simple provider of tires to a "provider of services for mobility" with an important range of telematic services and fleet management. This has made it possible to achieve customer fidelity.

Power by Hour

The reason why pay-per-use still leaves some big players perplexed can be found in the relative cost of the product, the purchasing cycles and again in the pool of existing and potential customers, as well as the costs of making the change.

In the mid-1980s Rolls Royce, followed by General Electric, introduced "Power by the Hour" in the market for reaction engines.[8]

With Power by the Hour, the customers – airlines or air travel operators – paid for actual running time and availability of reaction engines.

Although today this can hardly be called a "new" pay-by-use model, at the time it had the (great) merit of re-aligning prices more with use than with sales.

Power by the Hour was not actually a big challenge – indeed most leading companies managed to adopt the model in their various sectors – and the reasons for this are to be found precisely in the concentrated customer pool on a market that had been poorly served until shortly before.

Every market challenge is a new frontier.

Pay-per-use pricing tends to come as a revolution on those markets where the customer base can expand drastically, whilst the airline industry – with its regulations and other fairly high barriers to entry – does not grow so quickly as other, bigger sectors with many more players and lower entry barriers.

Another example of payment according to units of time – hours, in this case – comes from Zipcar, a U.S. car-sharing company: payment is made on the basis of the total number of hours the car is in use.[9] It is not rare for customers to pay a fixed quota. Nonetheless, this sum tends to be significantly lower than the purchasing price a customer would have to pay for the vehicle. In the case of Zipcar, clients pay an annual quota of $60 for accessing the entire fleet – with a definite plus effect due to the perception of an ample choice – whilst they pay up to $8 an hour for using the car.

Paying by Cubic Meter of Compressed Air

Even a company founded in 1873, the biggest in the country in terms of size, as well as being a world leader, can take advantage of the opportunity of new approaches to pricing and revolutionize its approach to monetization to consolidate its competitive edge.

This is the case of the Swedish firm Atlas Copco, leading producer of compressors.

With its new AIRPlan offer, in practice the company asks their customers: why not leave the equipment in the hands of Atlas Copco? With AIRPlan, the compressed air that is needed is obtained and payment is made on the basis of how much is consumed.[10]

In their presentation of this model Atlas Copco responds to an indirect, rhetorical question:

> *What is the difference to possessing your own compressors? Purchasing a compressed air plant makes a large impact on your fixed assets. As well as the cost of the investment, many other costs have to be factored in: administrative and capital costs, transport and installation, etc. With AIRPlan there is no need for any kind of asset to be purchased. All the costs of the*

compressed air are part of operating costs. And . . . freeing up cash for other investments could bring new business opportunities.

And so comes payment on the basis of cubic-meter consumption.

At the time when Atlas Copco started out on its path towards a shift in pricing, the trading mantra focused wholly on the evidence that equipment producers would be more competitive if they concentrated on producing the equipment and ignored all downstream activities such as contact with customers and assistance to distributors and retailers.

Nevertheless, Atlas Copco decided to concentrate on quality service and direct interaction with their customers instead of passing through distributors. This meant creating a direct network of sales staff and technical assistants who were to operate through a global Customer Center infrastructure and, in the long term, gradually convert indirect channels into direct ones: "We wanted to be sure we had our customer relations well in hand," stresses Ronnie Leten, ex-President of the Compressors Division of the Atlas Copco Group in one of his interviews telling the story of the company: "That way, on the side of the delivery chain we basically depended on collaboration with our suppliers, whilst our downstream business model was proving to be more or less vertically integrated with our customers. This 'close contact' with customers is a clear contrast to our competitors' approach, which had a less forward integrated business model and operated through distribution channels":

Once the branch's infrastructure became operational, the service business started to grow, driven by customer demand.

The customers demanded services and Atlas Copco responded!

But whilst, especially at the beginning, the demands were for simple, transactional assistance services, like everything else there was an evolution in demand which, in turn, encouraged the company to broaden its offer.

This is the virtuous cycle of good ideas which, like culture, knowledge and good practices, are assets that, when shared with many others, increase in value instead of diminishing.

Another aspect of the value gained through this type of monetization was perhaps more difficult to quantify but equally tangible.

Closer customer relations for Atlas Copco meant being constantly faced with the client's changing demands. In turn, on the company's side, regular contact implied that it was the first to anticipate demands for additional products or services. Therefore, in practice, intimate knowledge of the client – combined with ongoing innovation – stopped competitors from being able to intervene.[11] This is one of the many examples of how a perfect transition can be made from a product-centric to a customer-centric type of business, at the same time managing to consolidate the company's position through a "protection network" provided by its own competitive edge.

Payment by Weight

In air travel the price is traditionally set per person, though always differentiated according to age, status, or similar criteria.

The Polynesian company Samoa Air has proposed a completely different pricing standard.

Prices are set according to the passenger's weight: a fixed price is thus paid by kilogram, varying according to the length of the flight route.

Samoa Air's tickets vary from $1 to around $4.16 per kilogram. Passengers pay for their weight combined with that of their baggage.

For example, around $1 per kilogram of body weight is charged for a flight from Samoa to Faleolo.

Samoa has the world's third highest rate of overweight people, far above the United States, so that this pricing standard is a natural choice; even though some may see a discriminatory logic in it, in reality this is the logical application of personalization according to user.

Chris Langton, CEO of Samoa Air, was the keenest promoter of this pricing standard: "There are no extra costs in terms of excess baggage or anything else – it's just a kilo and a kilo is a kilo," adding that: "The smaller the airplane, the fewer variations can be accepted in terms of weight differences between passengers; what's more, people are generally bigger, broader and taller than 50 years ago."

With the new pricing model, some families with children would actually pay less for their tickets.

Logic speaks for the system.

After all, the passenger's weight and not their age or status is the cost factor here.

And, developing the logic behind all this. . . We calibrate systems according to the same measurements we create to control them. Why, then, if goods transportation is invoiced according to weight, shouldn't the same be true for people? This is what Samoa Air's managers asked themselves.

Langton also suggested that the move contributed to promoting awareness of the islands' state of health. They have one of the highest rates of obesity in the world: the 2021 UN Report shows that 84.7% of Samoa's population is overweight. Translated into numbers: only 31,000 out of a population of around 200,000 inhabitants are "normal."[12]

Be this as it may, for the moment this pricing model has merely been an experiment with a time limit, perhaps because of the discriminatory implications. In any case, what is happening now is that some American airlines ask passengers who are severely overweight to buy two tickets when a flight is full.

Adoption and Limitations

With technological progress, stronger connectivity through the Internet of Things and the potential increasingly smaller dimension of transactions – made possible, for example, by the *blockchain*, where bitcoins and crypto-currency are just the tip of the iceberg – the feasibility and desirability of pay-by-use pricing models will increase, above all on markets on which the products or services can be forwarded swiftly to customers.

Artificial intelligence and automatic learning, the increase in connectivity and integrated data analysis allow companies offering services and products to gain a far deeper knowledge of when, where, and how their customers use products and services.

The deeper knowledge generated can thus be analyzed for the purpose of further developing previous products or services to satisfy customer demands. In turn, customers benefit from more direct and personalized experiences than occur with products.

Market development of this sort represents a challenge for historical players who wish to protect their traditional customer base without compromising the flow of revenue and questioning basic assumptions as to what customers appreciate and how this (value) should be provided.

Typically, mature companies hesitate to adopt business models based on usage: they want to avoid the risk of cannibalizing the income generated by the initial purchase of their product. Furthermore they need to revise their sales-steering – different incentives will be required (not just sales-volume).

In traditional models the sales, support and distribution resources are also optimized for the anticipation of large-scale purchases, which does not favor pay-per-use models.

In addition, adopting a dynamic model based on usage risks compromising relations with existing customers who have purchased the product in advance. For these companies, which for years have had their roots in sales of high-cost assets and services to a restricted market, this model challenges the basic assumptions about who their customers are and what they need.

The so-called *incumbent* companies have been pioneers in this particular market segment. They have their own self-esteem, which they have built up over the years. As well as this, they have loyal customers, their brand is established on the market, over the years they have set up a network of relations with other companies and a logistics and commercial network. All this contributes to making them hostile to change and their knowledge of the sector almost represents a limit.

The questions that the management of these companies poses, however, are often the same; the scenarios change, the human being remains the same: "If our past success is due to the sale of complex products at a high price to big customers demanding ownership, why should we change our strategy and sell at lower prices, unpredictably, to smaller customers?"

For many companies the transition from a traditional model to a use-based one is not easy, even though the main product remains substantially the same. With dynamic pay-per-use models on offer for several types of products, both B2C and B2B customers are increasingly starting to expect wide and differentiated ranges of offer.

The adoption of pay-per-use models will be important for products with a high initial purchasing price on markets where customer use is dynamic, volatile, unpredictable. It will also be applied for expensive, technology-based products, or preferences that change rapidly – but is there anything that does not nowadays? – and whose use is irregular or cyclic.

In the B2C automobile or insurance businesses, or in upmarket fashion businesses, we are already experiencing the start of on-demand services in the field of infrastructures or less demanding products. Instead of automobiles, end customers can buy miles; work-specific insurance can be bought instead of annual up-front policies; and outfits acquired for a gala reception rather than for a lifetime. The same goes for the B2B sector, where companies are already counting on cloud-based products to sustain growth and variable demand.

There are, nonetheless, certain limitations in *pay-per-use*: low-cost physical assets in constant demand are the most resistant. One example is upmarket running shoes. A product like this will probably continue to be purchased in the traditional way because of the difficulty of delivering them on request, the rapid deterioration of the product and the (all in all fairly understandable) unwillingness to accept shared usage.

Another barrier to the adoption of pricing models oriented towards usage, like pay-per-wash, is the availability of the financial resources necessary (for example, for financing the creation of a base installed on dishwashers, which must be financed and/or insured). But although not all companies can afford the necessary financing, on payment of a commission a series of credit institutions are ready to offer support.

Although use-based pricing can make its bid by offering innovations in transactions in many sectors, there are some positive exceptions that confirm a rule: this happens in the reaction-engine industry, which has shifted without much disruption to a form of use-based pricing, partly because the entry barriers (both for customers and for suppliers) have prevented new players from entering the market.

Despite the limits and barriers that have been discussed, we can conclude by stating that companies operating in sectors that differ (even greatly) from one another benefit from the shift from traditional pricing models – based on ownership of the asset purchased – to approaches that place monetization of value at the center of their revenue modeling: the innovative nature and creativity of these pricing models, which exploit to the utmost the use of technology and digital solutions *in itinere* make it possible to create a competitive advantage by doing away with resistance to purchasing and attracting new customers.

Summary

Pay-per-use, or payment for usage, is payment to the supplier for a product or service according to its actual usage.

Compared to renting or leasing, which typically give the consumer complete rights of usage for a limited period of time, use-based payment closely links payment to the customer's patterns of usage.

This makes pay-per-use more attractive to consumers who do not use a product so frequently. Thus pay-per-use models make it possible to access quality resources without any significant capital outlay.

With the rapid growth in cloud computing and technological progress in general, as well as in data management, payment for use is spreading in a number of sectors. The Winterhalter case in the sector of domestic electrical products, Rolls Royce in that of aeronautic engines, Atlas Copco in that of compressed air and Zipcar in mobility are only a few of the numerous sectors in which it is being applied.

The introduction of pricing models based on payment for use may occur for different reasons: the need for greater flexibility, generation of cash flow, economic accessibility, customer satisfaction or in order to avoid the burden of ownership.

When correctly set up, they make it possible to remove the barriers to purchasing and monetize the value provided to the customer according to a precise target. Innovative businesses that adopt payment for use can benefit from important scale economies, even winning a substantial market share from players who continue to limit their offer to the sale of goods.

CHAPTER 3
SUBSCRIPTION PRICING

"The customer base is the new engine of growth."

Shanantanu Nayaren, CEO Adobe

Case History

Enormous metal dinosaurs at dawn. The sun's rays reflect on the presses and bounce off the outline of the screw and the chrome springs. Dust enters through the big windows, which sooner or later have to be cleaned.

As in the iconic photo published in 1932 in the *New York Herald Tribune*'s Sunday supplement, entitled *Lunch atop a Skyscraper* and taken at the Rockefeller Center in New York, 11 workmen eat lunch on a metal beam suspended a hundred or so meters above the city.

In this case, the images show the determination of human beings to move on despite the Great Depression that was happening at the time.

In the same way, to exit the dual financial and economic-ecological crisis at the start of this twenty-first century, another iconic image will be needed, as the key to an archetype of the future and of hope.

Let us imagine world leaders in the production of machine tools.

We sell printing machines – so-called sheet offset machines – bought by graphics companies the world over.

For over 170 years we've been selling these products at a high price.

But from one day to the next, aware of what's happening in the world, and not out of philanthropy but according to purely pragmatic thinking, we decide to change our pricing model.

After a process of concept creation and comparison, we move on to prototypes and hypotheses for monetary innovation.

There is no lack of suggestions. Next, we shall need an image to narrate our idea. This will come in due course; for the moment the practice foresees certain solutions.

The following offer will be made to our best customers: instead of buying expensive machines – we're talking an average of $2.5 million – a flat rate will be offered for printing a fixed number of sheets.

For $100,000 a month, the machine will be installed in the client's production plant. In this subscription, maintenance and management through big data are factored in. Paper, colors, lacquers, and also detergents and rubber pads for cleaning the press plates are included. If the customer produces more than a predetermined number, for example 30 million sheets, then the subscription can be extended.

What might be the iconic image for this campaign?

Men sitting on a beam, tightrope walkers, blue sky, and clouds. Rotogravure machines with newspapers pouring out of them. Printing presses, new Transformers, reassembling. In their new form as airplanes they take our material all over the world. To Paris. To the center of Caracas. The faces of kids in the slums of Nairobi. Jakarta submerged. Oceanix, the new island. City forests in Stockholm,

Rome, Delhi, New York: *No place is far away.* Claim on a black background.

This is not a movie but what actually happened in the Heidelberger Druckmaschinen company, with the "Heidelberg Subscription"[1] offer.

Media companies took the first step, followed by software companies.

Now, forms of subscription-based revenue are taking root in all sectors, all over the world.

Right now, encrypted codes, letters on a keyboard, cursors move figures, laughable figures, time-based consumption. You watch TV, decide to see a program right when you want to, sitting on the floor eating pizza, a picnic on the parquet in your own home, the kids knock over the beer but there's more in the fridge, sparkling water and lemon. Perhaps this is the image of a possible new life that's within reach, just that we never saw it.

The essential is invisible to the eye, says the Little Prince.

Time.

The supreme value.

After the 1800s had anticipated and overtaken it, the 1900s lost it (Proust), the two wars and then the 2000s accelerated, shattered and even dispersed it, atomic time, today perhaps we have realized that it's actually the only thing that counts. Space, real and digital, is the vehicle that travels through it.

Time that travels more rapidly upwards than downwards.

Sinusoidal curves that have altered our perception of linear time forever.

Today we know that all is one, recursive, the seasons have disappeared and we shall need to adapt more and more. The strongest will not win and Darwin, too, is left behind in the stream of time. Today the winner is the one who adapts first and better.

In the same way, instead of being subject to haphazard up-and-down trends, where more can be sold at certain times, followed by discounts in the times of "the lean kine" (with due respect to our Hindu brothers and sisters) and we have to tighten our belts, now, with subscriptions, even producers of machine-tools – or other durable assets – can benefit from a stable and in a way foreseeable income. To the same extent, this increases the value of the customer: the so-called *customer life time value.*

There is nothing more desirable, both for investors and for entrepreneurs, than observing "regular consumption and a consumption of services, against a fee," in other words the definition of a subscription.

This is why the Heidelberg Subscription is considered a win-win solution by the company offering it, both for the customer and for the producer: customers no longer have to sustain high fixed costs or deal with the stress of having to invest. Moreover, from Heidelberg's point of view, they even save money: in the case of a subscription, each sheet costs a mere 3 cents.

In the traditional model, counting the downtime too, it rises to 5 or even 6 cents a sheet.[2]

For Heidelberg, subscription is a panacea for the ills of the market – instability! – and makes it possible to sell more services and consumer materials, thus increasing the company's profit margins whilst making it independent of the haphazard up-and-down cycles.

Heidelberg are also convinced that they can manage their machines more efficiently than is possible for a single printing company: by managing a pool of machines counting 15,000 units thanks to their own cloud, an enormous quantity and quality of data is obtained with information that affects the optimal management of the printers.

In addition, with regard to "consumable goods," benefits from discounts for volume are obtained: the greater the number of sheets subscribed to, the higher is Heidelberg's revenue.

The management objectives of this German leader are thus to increase turnover and profit margins thanks to subscriptions.

Heidelberg's points of reference are technological players whose subscription models have generated great wealth.

Some 10 years ago in the software sector, we observed the spread of subscription offers by means of cloud technology.

Today, SaaS – Software as a Service – with a turnover of over $100 billion accounts for over a third of the global turnover of software producers. And they are growing constantly (according to estimates by the market research agency Gartner the growth rate is 20% per annum).[3]

By following this path, the CEO of Microsoft, Satya Nardella, managed to achieve the record for them amongst the world's most valuable companies.

Firms that succeed in establishing themselves thanks to *subscriptions* can achieve a growth rate that is five times higher than companies in the US S&P stock market sector: this is how Amazon, Salesforce, and SAP have become favorites with investors.

Analysis of Context

Subscriptions in B2C

The tendency by private users towards serial buying is rife in the field of digital consumerism and experienced even greater acceleration during the 2020–2022 Covid-19 pandemic. At first we were all taken aback, tuned in to a life hanging from a thread, a radio program, a news bulletin, the TV news, with bated breath, and then silence in the cities – Rome, New York, Moscow, Tokyo were asphalt deserts. Stags on the streets in Abruzzo, bears in city centers in Maine – nature took over human habitats, until the buzz returned to fill the world.

Inside our own homes, defeated by a minute being that obliged us to take a different view of time and social space and their echo in the individual, we found new energy where every effort seemed to be in vain, and managed to convert a microcosmos into a new space for imagination.

Perhaps it was partly thanks to this that Netflix or players in the streaming sector, like DAZN, Spotify, or Amazon Prime, experienced such rapid and significant development in this period of reconstruction.

The same is true for game developers, such as Blizzard (and the everlasting *World of Warcraft* saga), for the evergreen Sony and PlayStation, or the companies playing the Media stakes: from the *New York Times* to *Wall Street Journal*. Tim Cook, head of Apple since the departure of the Steve Jobs, has clear intentions: as well as music subscriptions, he intends to gain new flows of revenue thanks to films, games, and news on subscription.

Subscriptions for everyday products are already on offer: Procter & Gamble, for example, have offers of this sort for Pampers diapers and Gillette razor blades, like their keen competitor Dollar Shave Club. Amongst the pioneers of services regarding coffee machines and monthly deliveries of capsules we find Nestlé. The same goes for a wide variety of other goods categories: from the shoes on offer at $39.95 a month by JustFab[4] to pet food (dog food on offer at $18 a week from The Farmer's Dog[5]).

Tires can be sold on a pay-per-use basis, for example on a per mile driven for B2B customers, but also on a subscription basis for B2C customers. Zenises, a multinational tire company with headquarters in London and Dubai, that recently entered the *Guinness Book of World Records* by presenting the world's most expensive tire at $600,000 for a set of four tires, recently launched its tire subscription for B2C customers in Europe.

Zenises's new tire subscription business, Cartyzen, has been introduced by the company's CEO, Haarjeev Kandhari:

> *The service is currently available in Germany through Alzura X and its 600 partner stores nationwide. The model is based on small fees that users pay as long as they need the services: a monthly subscription of only €4.99 covers all new tyre related costs. Everything is done through the Cartyzen's online platform, which provides the services and solutions according to the customer, while collecting information about users' preferences in general. Cartyzen also guarantees coverage, i.e. tyre replacement in case of tread wear, punctures or incidental deterioration, regardless of the mileage already performed, in a measure designed to ensure customer satisfaction. Zenises is the world's first company to offer such tyre subscriptions, continuing its strategy of pioneering innovative tyre sales models. We are*

also the first tyre company to accept the cryptocurrency Bitcoin for all transactions.[6]

In Europe, families spend an average $130 a month, more or less 5% of the family budget on home purchases, like subscriptions to content such as[7] music, videos, software and games, as well as on deliveries of food such as fresh fruit and coffee, or beauty products.

The trend towards subscription buying is certainly nothing new, especially in the media sector, but the pandemic on the one hand and digitization on the other have generated truly significant acceleration of growth in this area.

In 2021, subscription – worldwide – was worth $700 billion, a sum that should triple by 2027, rising to $2,100+ billion worldwide.[8]

The pandemic has encouraged retailers such as Ocado and Morrisons in England to offer forms of subscription. In Italy, too, these are offered by companies like Barilla, Illy, or Scotti. Barilla, for example, has launched the *CucinaBarilla* package, which at a charge of around $40 a month, delivers a box directly to the customer's home containing nine "kits" at a time, chosen from amongst the many available recipes. Each kit yields at least two portions and contains the raw ingredients for the recipe in question. To prepare the recipe for the kit all you have to do is place the base in the right "intelligent" oven, also included in the subscription (provided free of charge to customers on loan for use and delivered directly to their home).[9]

Subscriptions are also gaining ground in other areas, such as the automobile sector.

BMW, Mercedes, and Porsche offer *all* subscriptions in some cities, or with certain dealers. Porsche started with *Porsche Passport*[10]: a subscription offered in North America, which includes insurance, taxes, maintenance and tire changes for a fee of $3,000 a month. The car manufacturer, whose unmistakable logo is the Stuttgart coat of arms with the horned horse at the center, also allows the model to be changed according to the subscriber's needs: a convertible in summer or a SUV in winter, for example. For the 911, in particular, in Europe a monthly fee of $1,899 is paid.

Volvo is amongst the most aggressive automobile producers in this field: the CARE subscriber program is destined over the next few years to generate as much as 50% of Volvo's turnover.[11]

Subscriptions clearly distinguish themselves from renting or leasing.

It isn't a matter of encouraging an expensive purchase using the weapon of deferred financing.

The objective here is rather to establish a lasting relationship with a customer with whom there would be no post-sales contact in most cases.

The connection, often a digital one, makes it possible to get to know the subscriber and offer added services in a targeted way. BMW, for example, offers a voice-commanded, digital personal assistant which looks for parking places, puts on favorite songs and reads emails.[12] The subscription to BMW's "Hey BMW" option costs up to $379, with truly interesting profit margins for the firm.

This makes it possible to increase turnover, even on the existing customer base and not only through the costly acquisition of new clients. On this theme, Shanantanu Nayaren, CEO of the pioneer of subscription Adobe, sums up the new mantra as follows: "The customer base is the new engine of growth."

B2B subscriptions

Subscriptions also offer great potential to manufacturing companies.

If products are sold, customers will buy: there will be an outlay of capital against the total acquisition of the asset in the customer's hands (problems included). If, instead, a service is given, the outlay is operational: this model implies an economic-financial passage and one of management and there is a shift from having ownership of the product to having *access* to it on a reactive, pay-by-use, basis. A subscription model such as the Heidelberg one goes further, by establishing a proactive, long-term relationship with the customer, through recurrent payment – programmed operational costs – which, in addition, outsource all the problems linked to the product or service to the producer, as summed up in Figure 3.1.

Many enterprises have started interconnecting their machinery or domestic electrical equipment through the Internet of Things (IoT), generating petabytes of data – millions of gigabytes. Nonetheless the use of the data is often limited. In the best of hypotheses they offer remote maintenance contracts, saving on a handful of technicians.

On the basis of a recent study by the management consulting company Horváth, only 5% of the companies considered prove to earn money with these services. A business model able to monetize the service is lacking. This is why a new approach is needed inside manufacturing companies: whilst in the analogical world the product can be perfected to suit all circumstances, even by creating offers and options the customer has not demanded, thanks to digital connection the actual use of a commodity can be measured down to the last detail. This creation of value is not the center of attention for many players in this sector. As it is in the case of Mann+Hummel, world leaders in air filters.

In 2018 on the US market, Mann+Hummel came onto the scene with the Senzit[13] subscription service.

Differentiating element	Product	Service	Subscription
Pricing Model	Ownership	Reactive, on demand	Proactive, prevents demand
Success factors	Be top of mind when needs occurs		Continuously improve experience
Client's point of view	I will solve my problem	You solve my problem solved	I don't want to have any problem
Offering focus	CapEx, i.e. capital expenditure	OpEx, i.e. occasional, operative expenditure	OpEx, i.e. programmed, operative expenditure
Data exchange	Single – during the sale	Several – on use basis	Continuous

Figure 3.1: Differences between product, service, and subscription

For $199 a month an intelligent sensor was provided for the air filter which, once installed on a harvester or digger, conveyed position, condition, running time, and filtering capacity directly to the *senzit. io* portal. For $20 a month fleet managers or building entrepreneurs also have the possibility of being informed directly on their smartphones when the filters are changed.

In this way Mann+Hummel do not limit themselves to merely selling products but provide a service for maximizing the availability of expensive machinery.

This service can be extended to the automatic delivery of spare parts or *in loco* maintenance, which may become a field to cultivate. Imagine a maintenance trip in a country location – cornfields and flowers budding in springtime, the company's white wine in the fridge just waiting for the filter-check round to finish, while the workers come to the end of their shift, fresh fruit is to be had straight off the tree and there's plenty of fresh air. What more do I want? (and yes, you'll have thought this, and no, it isn't hidden publicity, it's the demonstration of how much can be done through the evocative power of words, even unspoken ones, in the mind of the reader or the listener: persuasion, symbols, in the end human beings are no more than this, from *Sapiens* – 300,000 years ago now – shadows and stories, interpretations, and symbols, we're right back there, back to perception again).

But subscriptions can also require big financial investment, as in the case of Heidelberger Druckmaschinen.

Trumpf, leading producers of machine tools and lasers for industrial processing, have handed over the financing risk to their own bank, Trumpf Bank. At the end of the typical 36-month subscription, Trumpf takes back the machines and sells them on, second-hand.

Vissmann, traditional producer of heating and air-conditioning plants, has moved in the same direction: recently they launched their heating subscription,[14] promoting it with the slogan "Easy as your music subscription": for $106 a month a 10-year contract can be signed for a new heating system/service that includes mainte-nance, repairs, chimney sweeps, and gas supply.

The demand for this type of subscription is driven by the fact that for an outlay of $20,000, in the end the customer manages to obtain a new system by spending a little over $100 a month.

Landlords in particular favor this offer, as the monthly rate can be included in the rent paid each month by the tenant or deducted from tax returns as a running cost.

Five steps towards the successful launch of a subscription model

A subscription is therefore convenient because it provides services and allows assets, products and services to be used that would oth-erwise be impossible or too expensive to obtain. Last, but not less important, in a fluid age like the one we are living in it is *not* forever.

Here, then, is what has to be done, in just five steps.

1. *Plan the transition to subscription* Shifting from transaction-based management (or *una tantum*, I sell and take payment once only) to a subscription-based management model, with "recurrent" income, implies making several changes in a company: in terms of offer, pricing but also processes and IT systems. We must be aware that the *modus operandi* of our company will change; and a computerized system of self-analysis will certainly be required, calibrated to the offer of products, tracking of consumption, monitoring, and evaluation.

Metaphorically, there is a transition from the old world to the new. In the middle is an ocean, adventure and, in the same way, uncertainty and the need for a solid vessel.

The quicker the income mix migrates towards the subscription model, the more impetuous and revolutionary will be the path of transition. In this case it will be important right from the start to formulate a clear vision of the objectives guiding the route and plan the operational process accordingly.

For large companies even a rapid transition can take several years. This is why it is advisable to envisage an interim period in which new subscription offers are made alongside traditional ones, partly to reduce risks and in the meantime acquire useful experience in the pursuit of the new strategy.

2. *Basing offer on customers' needs* Success with subscriptions means 'calibrating' the structure of offer to the customers' needs.

Everything starts and ends with the customer. The certainty must exist that when offering a subscription this model of monetization is being exploited to provide customers with unique and important value that is fully in line with the way they wish to buy and consume the assets or services on offer, as shown in Figure 3.2.

Customer's needs	Customization, uncommon	Routine purchasing cycle		Variety, novelty, exploration
Value of subscription	Access to quality offers	Timely and convenient replenishment		Low risk, simple
Cases	• Emma & Chloé (jewels) • Gwynnie Bee (fashion) • Winc (wines)	• Dollar Shave Club (razors) • The Farmer's Dog (pet food) • Amazon Subscribe & Save (household goods)		• NatureBox (snacks) • Blue Apron (home cooking) • BirchBox (make-up)

Figure 3.2: Examples of customers' needs as drivers of the value of subscriptions

Source: Farknot Architect/Adobe Stock

As usual, it is not the answers we start out from. But the questions.

As Alejandro Jodorowski writes, "The answer is the question."

Human beings hide their wisdom in tales and stories and the questions reveal our mental processes and acquired knowledge. In this case, the company culture.

Where are our customers' *pain points*, then, and how can we help them out?

How do we segment the market?

What level or levels of subscription do we offer to whom and to what market segment?

Questions like these are not only fundamental for shedding light on today, on the basis of what we have acquired over time, they are also affected positively by all that we still want to achieve. They facilitate the future development we want to work towards today and the specific demands to be activated.

3. Determining the pricing of the offer Once the offer has been established, the price of the subscription has to be set. How? For example, by answering questions like: What do I make them pay for?

We might think of a time unit, or, as in the case of Heidelberger Druckmaschinen, of the number of sheets printed, or again number of users / sites / downloads and a lot more.

In addition, we shall have to think about the best way to establish this according to the model of subscription offered: is a flat rate best, or a hybrid model? And how should we differentiate price on the basis of readiness to pay? Right down to details like: what are the invoicing terms? For example, are discounts offered if advance payment is made?

These are only some of the questions that need clear answers right from the start.

4. Testing the offer Before launching a widespread subscription offer following a drastic transition, it is as well to verify the market response.

Trying out the subscription deal makes it possible to gather precious ideas for a massive launch, collecting feedback both on use and on how satisfied customers are with our offer. This makes it possible to project estimates of revenue flow and consolidate the relevant business plan.

5. Preparing the launch Last step: preparing to launch the offer.

Both from the point of view of communication to users – choosing the most suitable media, like Volvo who successfully activated social media to launch their Care subscription – and from that of internal communication: preparing sales staff to position the new pricing strategy alongside transactional business and providing the right incentives for allowing the offer to take off.

Summary

The mantra: "If you produce it, customers will come along," is a thing of the past which is not destined to return.

We have shifted from owning things to sharing them.

From buying assets to *acquiring* experiences.

We have shifted from having products to providing services – like finding ourselves in the midst of Erich Fromm's reflections on the enormous difference between being and "having" – or moving from solutions to revenue models based on results.

In the same way, companies are adapting their trading strategies and operations to become more customer-centric.

Rather than indirect sales by unit or user, they are selling directly to customers and on a recurring basis.

This is true both in the area of B2C and in that of B2B: we are experiencing a transition from an economy based on products and focusing on physical assets and transactions, to a fluid economy not bound to ownership.

In the first case the focus is on acquiring new customers with static, generic offers and making the sale through a single transaction before moving on to look for other customers.

In the second case, instead, the relationship with the customer is the hub of the business model: the purchasing experience is built up around individual customers, who experience a tailor-made service in line with their needs.

To be successful, five stages should be considered.

1. *Planning the transition to subscription*: shifting from management based on "recurring" income implies making a great many changes to the company: in terms of offer, pricing, but also processes and IT systems. The *modus operandi* must therefore be adapted.

2. *Basing offer on customers' needs*: success with subscriptions means 'calibrating' the structure of offer to the customers' needs.

3. *Determining the pricing of the offer*: once the offer has been established, the price of the subscription must be set.

4. *Testing the offer*: before the widespread launch of a subscription offer following a drastic transition, it is advisable to verify the market response.

5. *Preparing the launch*: planning communication and making the launch is the last step.

What counts in this approach to monetization is thus instant access, the result, not the ownership. Planned obsolescence is replaced by ongoing improvements able to satisfy growing expectations and engage customers in a lasting relationship.

The flexibility customers are offered also increases: packages based on volume, flat rates, long-term contracts are just some of the options clients are offered: this is the case of Salesforce, Zendesk, Uber, or Box – companies whose objective is to maintain a subscriber base, monitoring the use of the services that generate the recurrent income so highly appreciated by the stock market and by top management and constantly improving the service to obtain long-term customer loyalty.

CHAPTER 4
OUTCOME-BASED PRICING

"People don't want to buy a quarter-inch drill, they want a quarter-inch hole."

Theodore Levitt[1]

Case History

"How should a small comedy club react if the government raises the taxes on theater performances from 8% to 21%?"

This was the question that Teatreneu, an art venue in Barcelona, had to deal with when their audiences were more or less wiped out by tax increases in Spain.

The comedy club joined forces with the Cyranos McCann advertising agency.

Their challenge was to find a new strategy for increasing revenue after the sharp drop in ticket sales: briefly, 30% of income in just one year with an average 20% reduction in ticket prices and audiences shifting to alternative entertainment offers such as films.

The answer was to divide up human activity, in this case laughing, into measurable pieces of data and thus make it easier to evaluate in economic terms, too. It was the first case of the "pay-per-laugh" scheme offered to audiences for a comedy. The innovative payment scheme was made possible by using new technology: face recognition.

According to the programming parameters of face-recognition technology, it was possible to recognize a precise response, connecting it to various emotional states: laughing/happiness, crying/melancholy, surprise/fascination etc.

The pay-per-laugh app, initially installed on a tablet, was based in this case on a piece of software developed as a face tracker, or detector of facial expressions, capable of counting, listing, and generating statistics according to the number of laughs detected.

Every time it picked up a smile, the tablet took a photo and saved it. The technology, built in to tablets installed on the back of each armchair seat monitored the spectator.[2]

The agreement offered was simple and effective.

Entry is free.

> *If the show doesn't make you laugh, you don't pay.*
> *But if you laugh, you pay according to every laugh the actors get out of you.*

At the end of the show, spectators could check the laugh count, see the photos of every smile and even share them on social networks.

A laugh was priced at €0.30 up to a maximum cost of €24, corresponding to 80 laughs. This was the maximum parameter established and the limit couldn't be raised, so that the audience didn't have to try and control their impulses and laugh less in order to pay less, but were able to enjoy the show.

OUTCOME-BASED PRICING

From the 81st laugh onwards the company guaranteed the same fixed price, therefore the maximum paid was that fixed for the 80 laughs.

The pay-per-laugh app made its first public appearance at the Teatro Aquitània, in Barcelona.

When Teatreneu, in collaboration with the production company Canada, debuted with the comedy *Improshow*, the average ticket price had increased by €6 and the number of spectators by 35%.

Every pay-per-laugh show produced an overall counter value of €28,000 extra compared to normal box-office using the traditional payment system.

The system was copied by other theaters in Spain.

A mobile app was created as a payment system.

The first subscription based on number of laughs and not on number of shows was launched.

We are looking here at an exemplary case of pricing based on results, known as *outcome-* or *performance-based pricing*, or a revenue model aligned with the performance offered by those who provide a service or product. There are two elements that make this sort of monetization possible: the outcome-based revenue model and the technology.

The revenue model is based on the result of performance, in this case the entertainment: the laugh resulting from an entertaining theater performance is the visible result of the service offered. In terms of payment it is in the hands of the spectators: they are the ones who decide by means of their laughs how much the theater company will earn. If they don't laugh, no income will be generated. The risk is born entirely by the company: they must be certain of the quality of their performances and the value being provided to their customers. What is more, they must establish the pricing meter that is best suited to monetizing the value provided to customers, their audience. Finding the "perfect" pricing metrics capable of capturing 100% of the value provided to customers is no small matter but it is possible to get as near as possible to it. Using this approach, Teatraneu has managed to increase its revenue significantly; without it, the theater would have faced bankruptcy.

There are those who fear that some spectators may appreciate the show a lot but not laugh or that others might try and keep a straight face so as not to pay. But let's be honest – only an idiot would go and sit through two hours of comedy making every effort not to laugh.

The second aspect is the technological component: without the tablets installed on each seat, the software allowing face recognition, the "laugh count," the final billing, and the chance of sharing the experience on social media, it wouldn't even be possible to offer a pay-by-laugh formula.

Analysis of Context

Origin

Outcome-based pricing, too, has ancient origins.

Though not based on historical evidence, the story goes that a certain Chinese Emperor's personal physician was paid according to how many days the sovereign enjoyed good health.

The new, twenty-first century technologies make it easier and easier to base pricing metrics on performance: by combining digital platforms, machine learning, cloud computing, and the Internet of Things, the customers' conditions (health, too) will be monitored in order to produce more sophisticated solutions and give a better guide to their needs.

In the field of health, for example, we can imagine that it will be possible to measure the effects of drugs, medical devices, or certain services by means of sensors.

Prices could be set on the basis of real outcome. Of course in this case, too, the technically measured value must be translated into price units. This is basically no different from the general issue of benefits being expressed in prices.

Thus, using this model of pricing, customers pay on the basis of outcome and perceived value. And the more closely the pricing aligns with the value recognized by the customer, the more successful the business will be.

The risk linked to performance is born entirely by the company that provides the product or service. No result means no payment. The customers benefit from a reliable and predictable performance, otherwise they don't pay.

But what do we mean by result?

The starting point is to have a clear definition of the result in the organization.

The result is marked by three elements.

In order to be suitable as a basis for a model of monetization, a result must first of all be *important and substantial* to the customer. This may seem obvious but many companies neglect this point and focus on the characteristics of the product or service they have an intrinsic interest in or where they have a technological advantage, even when these characteristics are unimportant or just a "nice-to-have" that do not correspond to the customer's readiness to pay.

Secondly it must be *measurable*. The organization and its customers must agree on one or more parameters that best reflect the results, so as to be able to verify their actual outcome.

Lastly, it must be *independent*. Neither the company nor its customers, nor third parties must be able to tweak the result to their own advantage. This is the only way to have a result that is objectively suited to obtaining a recompense.

Now let us take a look at a series of applications of the outcome-based concept of pricing.

A click as a result

In the world of advertising it was always hard to quantify the impact of an ad: "Half the money I spend on advertising is wasted; the trouble is I don't know which half," admitted John Wanamaker, the retail magnate, over a century ago.

Since then, up to the advent of Internet, advertising continued to be sold on the web according to traditional pricing models based on exposure, such as the fixed rate or the "impression" pricing model (i.e. payment every time the ad is displayed on a website).

Nonetheless, over time far more innovative pricing models have emerged, based on the actions a user performs in response to an ad.

Today this type of model is becoming dominant and more popular than the more traditional ones.

The straw that broke the camel's back came from Procter & Gamble who some time ago negotiated an agreement with Yahoo! by which the portal charged the single ad on the basis of the number of clicks on it, hence the name "pay-per-click." So Yahoo! was only paid when a user clicked on the ad.

This happened on Google, too, so that pay-per-click has become the most widely used pricing model in paid search advertising.

Today Google declares to its customers: "You only pay if you get results, for example: a click on your website or direct calls."[3] This is the essence of Google's offer as summed up in their annual report: "Advertising based on cost per click, you are only charged when a user clicks on an ad on Google or . . . when s/he watches a commercial on YouTube" (Google purchased the video platform in 2006 for the, then, record sum of $1.65 billion).

The outcome is the center of the monetization model; if in the past a flat rate was charged hoping to catch the user's attention, now a charge is only made if the user has actually seen the ad.

Goggle is going even further: now there is the option to pay for conversions, rather than for clicks. In *pay-per-conversion* – also known as *pay-per-action* – the advertiser pays for conversions, that is, only when customers move from the banner to their website and make a purchase.[4]

Thanks to these monetization models, in 2020 Google succeeded in earning a record figure for advertising revenue: $147 billion.[5]

Kilowatt hours as a result

Those who invest in wind power have a sole objective: to produce energy. If you're a supplier, why not charge, then, on the basis of the energy produced?

According to this logic, Enercon, leading supplier of wind turbines, apply a fairly innovative sort of price metrics. The charge is calculated on the basis of the annual energy output actually achieved by the wind turbines.

The company will only be paid when their customers produce energy. In periods of high winds with high outputs, the customers will pay more; in periods of low wind, with correspondingly low energy output, they pay less.

The novelty is that Enercon participates in their customers' business risk. Enercon actually assumes a substantial part of the risk.[6]

The contract stipulated is called *EPC – Enercon Partner Concept –* and includes maintenance, assistance, and repairs. The customers pay a minimum quota according to the type of turbine used. This minimum includes the following services: regular maintenance, guarantee of availability, repairs including spare parts, transport, and remote monitoring 24 hours a day.

To keep costs as low as possible, especially in the first five years of operation, Enercom also takes responsibility for half the fees payable to the EPC over this period. Not until the sixth year of operation, does the customer pay the entire fee by applying a simple formula: charge = kWh generated × price per kWh.

This innovative service and the pricing offer are obviously welcomed by customers. Around 90% of them sign a contract according to the EPC proposal. An important prerequisite for the success of the concept is that Enercon is able to measure the performance of the wind turbine itself and therefore any manipulation by the customer is out of the question.

Hours of light as a result

What do Ikea, Walmart, Aldi, and Apcoa have in common?

All these companies offer their customers parking, some of it under cover, which must be illuminated, at least for part of the day.

There were several companies that looked after these parking lots in the traditional manner: they sold spare parts, such as new light bulbs, by the unit, charging an hourly rate for any sort of maintenance service.

It is easy to imagine that this is not a very differentiated market and a highly competitive one, on which offers from the various providers are easy to compare and suppliers come under a lot of pressure. In most cases whoever offers the best price wins the competition. Simple.

But it isn't the actual light provided by the bulbs that Ikea appreciates, so much as the confidence their customers get from a well-lit parking lot.

If one of the lights in the parking lot was faulty, the supermarket called a technician to replace it.

If this operation took too long, the customers complained that the parking lot was unsafe and in the end perhaps they went to shop (and park) somewhere else, causing the company to lose income.

If. If.

Whilst in sci-fi, the "what if" is a necessary condition and sufficient to create those imaginary worlds of science fiction, in reality "ifs" don't produce results.

After talking to the managers of a chain of supermarkets he was serving, a forward-looking supplier realized the true value of his offer and recognized an opportunity, as well as a need, to change the rules of the game.

He went back to the managers with an idea and a pricing model based on outcome: payment based on the number of hours the parking lot was fully illuminated.

If a bulb breaks, the chain of supermarkets doesn't have to pay.

Of course the bulbs are always in perfect condition: the suppliers have their maintenance team, which carries out regular checks. This results in a significant reduction in costs, because the emergency staff is reduced and the supermarket is a happier customer, which uses its "own" ability to guarantee a safe, well-lit parking lot in its marketing initiatives.

And suppose, instead of the parking lots, it was your company that didn't have to buy light bulbs or lights and could simply pay for the light it consumed in a responsible way?

This means no accessories, bulbs, LED . . . better, there is no actual need to possess any sort of lighting product! You don't even have to think about it.

This is the idea behind the monetization model conceived by the CEO of Philips, Frans van Houten, who conceives of lighting in a completely new way, recognizing the customers' need for so many hours of light in their offices: it is not the product that is of interest but the result.

Customers "just" want to buy the light, nothing else.

What's being sold here is the outcome, the light, and no longer the product.

The company's customers thus pay Philips a flat rate to manage an entire lighting service (planning, equipment, installation, maintenance, and updating) and thus the light consumed (the result).

A tailor-made system makes it possible to save on the initial costs associated with the installation of energy-saving lighting.

By planning a long-lasting service, rather than a "fit and forget" approach, lighting is provided in the most efficient and economic way possible – and this encourages the use of energy-saving lighting. There is another *green* benefit from the operation: at the end of the contract, the products can be re-used, thus cutting down on waste.

The Washington D.C. subway was one of the first to subscribe to this model, together with the UK's National Union of Students and Rau Architects in the Netherlands: "We ended up by creating a minimalist lighting plan that used the building's natural sunlight as far as possible, so as to avoid wasting materials or energy," explained the heads of Rau: "A combined system of sensors and controllers also gave us a hand in helping to reduce energy use to an absolute minimum by regulating or turning on artificial lighting in response to the movement or presence of natural light."

On the other hand, from a business point of view there has always been a problem with LEDs: how do you make money when your product lasts decades? With the arrival on the market of more efficient technologies, Philips realized that they could sell this solution to their customers.

In 2014 *Fortune* nominated the Philips CEO, Frans van Houten, one of the 25 best "eco-innovators" in the world, recognizing him as a pioneer of this innovation.[7]

Broken rocks as a result

One-stop solutions can mean more benefits for customers in terms of greater safety and efficiency, and can revolutionize sectors which have been static for decades in the area of monetization.

This is the case of the commercial dynamite used in excavation work.

Up until yesterday what was the pricing model used? Simply the price per stick plus any services.

The Australian company Orica, world leader in the production of commercial explosives and sanding systems, has changed this old rule and now offers quarry operators a single solution. Not only do Orica provide the commercial explosives on the basis of the quality of the "broken rock" or degree to which the rock is fragmented, but they also analyze the rock and carry out drilling or blasting operations. In this system model, Orica provides customers with the crushed stone and charges them by the ton.

The result is what is called "rock on ground," where the dimensions of the rock the explosion generates are strongly correlated to its value to the customer. The smaller the fragments of rock are, the more quickly and easily will the excavating process be carried out.

Since this is a personalized solution, the prices are less comparable and the customers' income increases, as well as the efficiency and safety. Customers no longer have to concern themselves with the sanding process.

Outcome: it gets more difficult to change to another supplier.

Thanks to the new, digital BlastIQ program, Orica claims to "be in a position to provide predictable and sustainable improvements that can reduce the overall costs of excavation and blasting and raise productivity and safety, also ensuring . . . that customers take better and quicker decisions, which produce better results in their operations."

This is how the company has shifted from selling explosives that make a hole, to offering an integrated solution supported by data linked to blasting.

An analysis of the customers' data has made it possible to identify factors and schemes that affect the blast process. And so Orica even offers guaranteed results, within certain margins, which enable the effects of the explosions to be predicted, quantified, and monitored.

Operators of quarries or mines can therefore take targeted decisions on how to run their projects, with savings on time and money that would have been inconceivable before this pricing model came into operation.

Health as a result

You pay to be cured but what happens if you aren't cured?

We're used to buying drugs or treatment and paying a price independently of whether we actually recover or not.

Johnson & Johnson was one of the first companies to suggest a result-based pricing model in the field of oncology in England.

If the anti-cancer treatment doesn't prove effective, patients are reimbursed the entire sum of money spent on the treatment.

Others have moved in the same direction.

Roche, the Swiss pharmaceutical multi-national suggests personalized systems of reimbursement that make a clear break with the tradition of charging for a pill or other treatment, in other words the *proprietary model* inherited in this sector.

By using this new model, instead, Roche acknowledges that the effects of drugs can vary according to indications, that is, the patient's specific condition, combination with other drugs, and the response; in this way customers adapt to the new reality. In what Roche calls *pay-for-response*, the charge is based on the patient's response to their treatment with a specific pharmaceutical product over a determined period of time.[8]

Briefly, that patient signs a contract. The company agrees to refund the price paid if the treatment does not result in a successful outcome. Either directly or indirectly through intermediary partners.

In 2017 the pharmaceutical company Amgen and the insurance company Harvard Pilgrim came to an agreement of this type: Harvard Pilgrim benefits from a discount when a patient treated with Amgen's drug Repatha (which reduces the risk of heart attack by lowering levels of cholesterol) does not register significant improvements.

Medtronic: "Signed almost 1,000 contracts requiring the company to reimburse hospitals for the costs sustained, if its star antibacterial Tyrx failed to prevent infections in patients who received heart transplants," stressed Omar Ishrak, at that time CEO of the company.[9] The company also has a reimbursement agreement with the Aetna insurance group if diabetics fail to improve when they shift to Medtronic treatments. Other result-based contracts are being evaluated.

But the Minneapolis med-tech giant is not alone.

GE Healthcare and Philips are amongst other producers who link payment to real results. The shift to contracts and partnerships based on monetization linked to results is a natural evolution of the more widely ranging shift to result-based treatment.

"Med-tech companies are actively looking for opportunities to commit in new ways with hospitals and physicians, and looking for methods of sharing the risk these sorts of suppliers are running with their new models of payment, as well as sharing the benefits," said Don May, Executive Vice President of payment and healthcare delivery policy for AdvaMed.

GE has launched an important initiative for incorporating and interconnecting digital sensors in its medical equipment, aeronautical engines, power turbines and other equipment. And this is only the start. GE is a multinational that has started out on the route of digital transformation, offering result-based services, where customers only pay for output generated by GE on the basis of KPIs, or agreed performance indicators. This transformation has yielded its fruits. The company is now generating $2 billion a year from results-based services, which only its medical services unit[10] provides.

These opportunities have a wide range of application: they range from agreements on pricing, like Medtronic's Tyrx, where there is a price for achieving quality metrics and another if they are not achieved, to the other examples quoted, all of which are linked to results, even though some in particular combine a traditional, transactional pricing model resulting in the patient possessing the drug or the treatment against payment, with a promise of reimbursement if the expected outcome should not be obtained.

Insured risk as a result

Some health-promoting activities should receive incentives through less expensive insurance premiums.

There are many possible, simultaneous applications for the use of new pricing metrics in the health sector. Here measurements can be carried out by smart watches, sensors integrated into special bracelets, or other forms of remote diagnostics.

The British health insurance firm AIG Direct uses the body mass index (BMI) as the basis for calculating the monthly rate. Exceptions are only made in special cases where the person insured practices a particularly high level of sport, in some cases even competitive, so that the BMI would prove to be "distorted" by the larger muscles developed.

Moreover, the pricing incentives can be used to reward desirable behavior and make the undesirable subject to sanctions. It is obviously up to us to decide what these are in the framework of a wider context, company by company, according to a broader reflection on the objectives to be achieved on a short- medium- or long-term basis.

Transformation of pricing and of the company

This type of monetization can be attractive both to the buyer and to the supplier. In practice it simplifies the buyer's life and if customers fail to receive the results guaranteed, they simply don't pay and in some cases penalties can even be applied.

On the other hand, sellers assume the risk but create value by solving the customer's complexity, setting the price of the service according to the value created.

This way of doing business can improve customer relations, being integrated somehow into the contract together with profitable activities regarding long-term assistance and maintenance.

This is why some companies are starting to talk about *outcome-as-a-service*: the result in terms of a service. This new approach to monetization requires a different relationship with the customer compared to the one that existed in transactional pricing.

The relationship begins with identifying the problem the customer is really trying to solve. It is essential to listen to what customers say, especially if they don't know exactly what they want and/or what the company can offer. What service to offer after listening (a rare commodity in our times)... does the customer want the guarantee of a 24-hour service, 7 days a week, or is the objective to maximize income? Since there may be several scenarios implying different operational protocols and levels of performance risk, making the customer's expectations clear is absolutely fundamental.

Result, as a model of monetization, also implies reporting. Continuous communication with the customer is essential and should be defined in the contract. It often involves several divisions: those responsible for loans, financial information, right down to emissions data.

Of course, any failure in performance becomes critical, or rather inevitable, both for documenting and for rectifying.

If it is not correctly managed, this failure may create a greater financial risk as well as damaging customer-supplier relations.

To sell successfully and provide results, the generators must reflect on (and in some cases restructure) the way they do business, from their initial marketing right down to delivery.

The sales process generally requires a dialogue between several parties for buyers and sellers alike. Whilst the profit margins are often considerably higher in this model, performance risks are, too.

Sales staff must fully understand the results they are selling and guarantee the costs of delivery to the whole organization, as well as the risks. The old sales model – winning a contract and pressing on – with the delivery phase at the end (the fatal responsibility) being someone else's affair, is no longer possible.

Integrated teams must collaborate, from the process of price setting right down to the actual delivery, concentrating on the service provided to the customer.

All aspects of the service must be defined and appropriately priced. The risk of the service failing, with downtime that can easily amount to millions of dollars in a relatively short time, now becomes the responsibility of the provider of solutions.

These risks must be evaluated and accurately factored into the product.

Suppliers must establish communication and feedback cycles, so that potential for failure is reduced to the minimum, at the same time ensuring that lessons learnt are incorporated into the organization. Programmed incentives must be adjusted so that entire teams are rewarded for performance and the creation of value.

From the customer's point of view, this must appear to be a continuous flow. Internally that means clearly assigning the supplier's responsibilities and developing the capacity for learning and adapting rapidly. If we think about it, in the end this is no different from all that goes with the concept of "evolution."

Summary

Companies in the widest variety of sectors in different nations and continents have left transactional revenue models behind them, to base their monetization on models that reflect value to the customer, charging according to results.

To be one, the result must have at least three characteristics: it must be important and substantial to the customer, measurable, and lastly, independent.

Examples of results are laughs in the entertainment sector, clicks in advertising, kWh in energy supply, and health in the medical sector, to mention only a few.

Both buyer and seller obtain advantages.

The buyer's life is simplified: if customers don't receive the results guaranteed, then they don't pay.

On the other hand, the seller assumes the risk, but sets the price of the service on the basis of the value created and the latter is fully monetized.

Outcome-based pricing models quantify and measure the result to the customer, thus the value, taking advantage of innovation, data, new technologies, and experience.

The creation of these capacities, together with increasing digitization, can represent a critical and demanding cultural change for companies.

The increase in profit margins and creation of competitive advantages linked to the model of monetization, or the longer contracts, make up for the risk assumed and create opportunities for consolidating customer relations, providing clients with the results and services they "really" desire.

CHAPTER 5
PSYCHOLOGICAL PRICING

"All our knowledge originates in our perceptions."

Leonardo Da Vinci

People's behavior does not depend only on the value of the goods and services available and on their respective prices, but also and mainly on the perceptions people have of things.

> **perception** n. [from Lat. *perceptio -onis*, deriv. of *percipĕre* «percepire», past. part. *perceptus*]. – 1. a. The act of perceiving, that is, of becoming aware of a reality considered external, through sensorial stimuli, analyzed and interpreted through psychological, intellectual processes.[1]

Many authors have written about perception: Aldous Huxley, Pavel Aleksandrovič Florenskij, Ernst Mach and the Gestalt theorists. Moreover, "We do not see things as they are; we see things as we are ourselves", recited the *Talmud*, the sacred Babylonian text; a statement then taken up by famous and illustrious figures, such as the German philosopher Immanuel Kant and the Swiss psychoanalyst Carl Gustav Jung or, again, the French writer Anaïs Nin.

Case History

Behavioral pricing takes into account this concept and acknowledges that customers may also behave irrationally.

A good example of irrational behavior? Here's one.[2]

The smell of saltwater in the air, stretched out on a towel, eyes closed, hands sunk into the sand, the skin burning although you've smothered it in sunscreen, it's the "perfect moment" – you know, you *feel* it – as you're lying there on an enchanting Caribbean beach (Rimini will do, too).

There's not a cloud in the sky and it's a wonderful, hot day.

In fact in the last couple of hours all you've done is think (obsessively) of how you'd love an ice-cold bottle of your favorite beer.

Like last summer, when you were on a diet – which, at the tenth, extra (obligatory) session, just before the bikini test, they tell us is called an "eating régime" – it was roast chicken and cakes floating around in your head while you slept, this year it's the "ice-cold" fantasy and the best way to cool down on the beach, when just a moment ago you were reading about the latest arson attacks and the climate crisis. You shake your head, and for a second you'd like to stop thinking. Just enjoy the well-deserved rest in silence. No thinking, *please.*

"Nothing will go well," you think; it's mathematics, it's the tenth *Murphy's Law.*

Instead, we should all know: the Universe is there ready to give us the lie and demonstrate to humankind once again, if it were necessary, that we know nothing. And even when we do know, we mystify, altering meanings to our (dis)advantage.

The same *Murphy's Law* doesn't actually say that things are bound to go badly. This is an interpretation. *Our* interpretation, as typical human beings; it says that things will not go as expected and this just means that the world is far bigger than we are, with our limited ability to make it standard by using mathematical equations and pertinent, or what are presumed to be rational, decisions.

In fact, in the precise instant you thought, "How I'd love an ice-cold beer," despite your most victimistic side emerging, a friend gets up to go and make a telephone call and says, "If you like, I'll bring you back a beer," so that you're about to burst into tears or hug them and you'd do it except you're sweating too much, even for your friend.

Your friend looks at you, gives a hint of a smile, but stands there, the last little gift inherited from the pandemic being "the right distance," which will perhaps never be in harmony with physical empathy; and you discover that the only place anywhere near where your friend is going and where they sell your beer is – of course – a luxury hotel (!).

"The beer might be expensive," insists the little snake, asking you how much you're prepared to pay.

He or she – but in our case "the bastard" is him, the friend you have in mind now, in *this* precise instant – presents you with a dilemma: he's read too much and is an incurable crossword fiend so, not yet satisfied, he comes out with the riddle of the three doors or *Monty Hall* solution, he mutters about non-counterintuitive solutions (!) and makes the following proposal: he'll buy the beer if it costs the same or less than you tell him; if, instead, the price of the beer is higher than you decide, he won't get it.

Apart from reviewing your friendship, what's the solution, since it's not possible for you to negotiate with the bartender? What price are you going to tell your friend?

Case History – *Altered Version*

Now imagine it again – the same scenario.

You're still stretched out on the beach on a hot day. For a couple of hours you've been thinking intensely how much you'd like an ice-cold bottle of your favorite beer. A friend stands up to go and make a phone call and offers to bring you one from the only place in the vicinity where they sell it, that is, a rundown little drugstore.

Right!

You see? The unconscious is already at work.

Possible alternatives. Identical scenarios. Just one small change. Let's see what sort of impact it has on the story.

Still smiling, the friend asks you how much you're prepared to pay for the beer.

What price will you tell him in this case?

Fine.

These two scenarios were presented to a fairly large sample group and on average they tended to name double the price for the beer from the resort, compared to the one from the drugstore.

From an economic point of view this is not rational and not what the theory of *homo oeconomicus* would expect: the beer is the same, the temperature on the beach is the same and it's not even a case of a direct buying experience from two different sales points, since your friend is buying the beer just for you.

Obviously, psychological factors impact strongly on readiness to pay, which is not only a result of the value that is received from the product. This recognition is the basis of behavioral economics: customers do not always act rationally.

The pricing tactics that play on the "dark side," the customer's irrational side, are summed up in the term *behavioral pricing*.

Analysis of Context

The nine behavioral pricing rules that make a difference

In behavioral pricing practice we find certain basic rules that can help companies to monetize the value delivered to the customer in a targeted manner. We have introduced them in many companies and they have instantly yielded clear results, partly combined with one another. Let us take a look at some of them.

1. Contextualizing value with a "price anchor"

Brothers Matt and Harry manage a shop in New York. They're pleasant, likeable, shrewd guys. Matt is the salesman – brown-eyed with a blond fringe; Harry is the tailor – pacific, careful, with a melancholy expression.

They sell suits – elegant, sober tweeds in Shetland wool. The counter running down the middle of the shop is made of wood and wrought iron, the image of past times with ladies in long skirts and gentlemen with waxed moustaches, when Europe was dancing the waltz beneath precious Bohemian candelabras in the *Ballhaus* before the war, or kilts in red and green tartan lying on the chairs – memories of times past.

In fact, Matt's and Harry's family is of Scottish origin on their father's side.

A gentle silence pervades the shop, particles of dust dance light-heartedly in the air, the sun's rays play on the objects there: scissors lying on the counter awaiting the next tailor-made suit, a pair of men's shoes, pieces of cotton on the partitions. Then a bell rings and a wooden door with glass panels lets the daylight in. The customer enters: "Good morning." He looks around, at the stairs up to the second floor where the hats and other accessories are. He takes a walk around and stops in front of a jacket.

When Matt senses that the customer likes the jacket, he winks at his brother. Harry goes downstairs to fetch some fabric and Matt deals with the customer, pretending to be a bit deaf.

When the customer asks the price of the suit, Matt shouts down to the basement, "Harry, how much is this suit?" Harry replies that he can't come up and look right now but if it's the one he thinks, "It's $92." Matt pretends not to hear. "How much?" he asks again. "92," Harry repeats from down below, his voice echoing as though underwater. So Matt turns to the customer and, with his best smile, says: "That's $42, thank you". The customer doesn't hesitate for one second, pays the price and vanishes.

What does this story teach us?

That the customer has fallen into Matt's and Harry's trap. They're Scotsmen and so shrewd, they could sell fire to the devil!

But it's also true that the customer didn't even bother to check the quality of the suit. . . He didn't even compare prices, as the game theory of *homo oeconomicus* would have expected.

Contrary to what we presume, buying decisions of this type happen frequently and rather than being based on pondered comparisons, prove to be the fruit of impulsive, and not infrequently irrational choices.

People evaluate prices as high or low on the basis of their mood of the moment.

For example, in a test people like the same wine "better" if the price is higher. This may seem absurd but let's just remember the beer example!

This is why some companies have been studying their customers' behavior for years, so as to understand how to maximize turnover and profits by means of an optimal pricing strategy. Innovative methodologies for measuring readiness to pay or predicting reactions to promotional offers or markdowns come into this area.

Another famous example of pricing is that of *The Economist*.[3]

A panel of readers is divided into two groups to carry out a test on pricing.

Group A is presented with two options: an online subscription for $59 or a subscription both online and in print for $125.

Group B was instead presented with three options: an online subscription for $59, a subscription exclusively on paper for $125 and a subscription both online and in print for the same sum of money.

The difference in the offers was therefore only the price of the subscription for the printed copy.

The effect of this pricing was to make the package online plus print subscription seem very reasonable, with the online subscription looking more or less like a free gift.

The print-plus-online package was chosen by as many as 84% of the readers in group B, whilst only 32% of group A opted for this package (see Figure 5.1[4]).

Once more, these two examples demonstrate the power of pricing.

By influencing price perception, a monetary reference point for orienting choice is fixed in the customer. If the monetary reference point is raised, inevitably higher prices will be obtained. The anchoring thus allows those who "set down" the price anchor to establish the reference point in their own favor. How?

By proceeding to influence the customer's readiness to pay.

2. Removing resistance to buying by using the printer/cartridge model

Our research, regarding both B2C and B2B customers, has shown that one of the main barriers to purchasing is the cost of the initial

The effect of price anchoring in magazine subscriptions

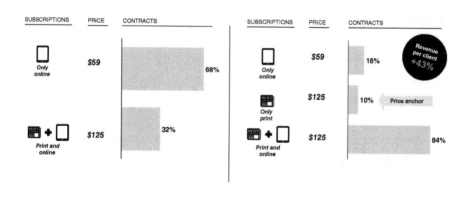

Figure 5.1: The effect of price anchoring on customers' preferences

outlay: although the sum of costs throughout the whole life cycle of a product is what must be considered, the initial outlay still remains the greatest obstacle to be overcome when selling.

Let us suppose we're a company that produces printers and that, as well as selling the main product line, we also sell cartridges.

We are in the phase of launching a new printer on the market with special *ink* cartridges.

Let's also assume that, as well as the printer, our target requires a cartridge every month. In terms of pricing the company's marketing team suggests two pricing models: the first model foresees a price of $510 and $20 per cartridge a month, whilst the second model foresees a price of $150 and $50 per cartridge a month.

Although a rational customer is indifferent to the two options, because over a 12-month period the overall price of the printer and cartridges is the same (i.e., $750), our research shows that the second pricing model wins – because the initial outlay required is lower. Thus, the customer believes that they somehow make a saving, whilst it is simply a matter of cost distribution, which is substantially spread over several months.

It's a matter of perception of time and of saving/investment.

The same goes for razors and razor blades, as well as for coffee machines and capsules, right down to elevator trucks which are sold with negative profit margins so as to then make money on post-sales services and spare parts. All in all, correlated products.

The motive, in both B2C and B2B contexts, is that from a psychological point of view the price paid at the moment of purchase has a clearly higher impact on the cost accumulated during use. In technical terms this is *cost of ownership*.

This is why, in establishing a pricing strategy, a customer is often acquired by means of a reasonably low initial price, followed by considerable variable costs. This is also the essence of customer life-cycle management: several companies, for example in the field of IT, sell an initial base version of their product, so as to later sell advanced versions by means of so-called upselling. Companies capable of offering and selling complementary products, such as razor blades, capsules, or spare parts will win out if they follow this pricing model.

3. Taking advantage of maximum readiness to pay by means of "threshold pricing"

In view of the fact that the subconscious strongly influences buying behavior, in determining prices an answer must be found to the question: how is the price perceived by the customer? And what reaction does the price produce in their brain?

Students of behavioral sciences and economists have been looking for the answers to these questions for a long time.

A price of $1.99 is associated with spending $1, rather than $2 – this we know. But why? The phenomenon is explained by a question of "numerical cognition": people tend to evaluate numbers with several decimal places by arranging them on a mental line. What is more, people read according to the interpretation of Arabic numerals – the numbers we use every day and which the decimal system merely perfects – that is, skimming prices reading from left to right. In practice this means that prices such as the price of gas, for example, often finish with a 9: the price for a liter of gas at $1.799 adds up to a total of $107.94 for a full tank of 60 liters. At a price of $1.8 the same full tank would instead cost $108. Despite the savings only amounting to 6 cents, drivers will nonetheless be ready to drive round town – basically consuming the 6-cent difference – until they find an apparently cheaper gas station that allows them to save money!

Stopping below the "psychological threshold" makes it possible to manipulate preferences even when the advantage offered is negligible. We have seen that in the end the customer wastes time (!) and money looking for what seems to be the cheapest gas station.

It should be added that customers often have a "price threshold" in mind: this implies that the perceived difference between $99.99 or $99 and $100 is far more than one cent or one dollar.

Over time, many market research studies have identified the $50 threshold, pricing at $49.99 rather than $51, so as to win a certain percentage of customers not ready to pay $51, but willing to meet the threshold price.

This may also mean raising the price without losing sales volume or damaging the company's pricing image.

For example, if a well-known producer of chewing gum offering a packet at 92 cents against a threshold price of $1, were to increase the price to the limit of the threshold, that is, 99 cents, they would manage to make 7 cents on every single packet sold without doing any harm to their pricing image. And according to the law of numbers, we know that with considerable sales volumes, the 7 cents can easily amount to a total profit of millions of dollars in the space of only a couple of years.[5]

4. Facilitating the customer's choices by means of the compromise effect

A centrally located wine bar. Bottles and chocolate. Main bar showcases regional, Italian, and international wines. Let's suppose we offer two bottles of wine to a panel of customers: a relatively expensive bottle is priced $50, another instead seems less expensive, at a price of $10.

The question is: how do I manage to pilot the consumer's choice, greatly increasing purchases of wine without lowering prices or making promotional offers?

The answer is: by using the compromise effect.

The compromise effect suggests that a product will be more likely to be chosen from a group when its attributes are not located at the extremes of the range of choice: in this case the best wine with the highest price or (in the spirit of *aut aut*, either/or) the one with the lowest price.

The compromise effect is one of conjugation – in fact, all that has to be done is change the "or" to a more inclusive, and above all alternative and easier "and."

It will be sufficient to introduce a third offer: for example a bottle at $30, and in this way a bigger turnover bringing greater profits will be encouraged.

Although there are always customers who are more price sensitive and will go for the $10 wine, and customers oriented towards more prestigious products who will choose the $50 one, most customers will be grateful to find a bottle at $30, and will opt for the intermediate choice: the one that is psychologically perceived as the "right compromise" between price and quality.

5. Using pricing as an indication of quality

Delvaux, producers of upmarket handbags, have succeeded in obtaining an image comparable to Louis Vuitton, thanks precisely to a sharp increase in their prices.

The same goes for the whisky makers Chivas Regal, who have created an elegant label and increased their prices by 20%.

In both cases a considerable increase in sales was obtained whilst profits grew out of all proportion. This is due to the fact that price is an indicator of the quality of a product or service: high prices stand for high quality, whilst low prices stand for lower quality.

It is mainly customers who are not very familiar with a product that are easily influenced.

Those who are unfamiliar with the quality and price of a product look for a parameter to guide them: they therefore associate high prices with better performance. This tendency has been confirmed by a study of some medical products.[6]

Considerable price increases without corresponding improvements in performance or perceived value are, nonetheless, risky and unadvisable. Pricing is a strong indicator and should be correctly used: in case of doubt, it is always better to price upwards – lowering the price whilst starting out from the perception of high quality is always simpler than raising the price when starting out from the perception of a low quality: price ratio.

6. Creating scarcity to encourage sales

Impulse buying is obtained by artificially created scarcity.

An experiment carried out in an American supermarket in Sioux City confirms this: Campbell's soup – the brand immortalized by Andy Warhol, who in 1962 produced 32 polymer canvases each with 32 symmetrically arranged cans of soup on it, consisting of all the recipes then on the market – was offered at a discount. On certain days in Sioux City a poster announced: "Maximum 12 soups per person." On other days, the same poster read, instead: "No limit per person." The result was that when the limit was set, customers bought an average of seven soups: double what was purchased on days with no rationing.

Now let's imagine we're in a clothing store. We're lucky; our favorite jeans – yes, *them* – are available and next week they will be marked down 10%. But as we are looking for the right size, we come across a disturbing element: only two pairs of the right size are available. What should we do then? The same old question: buy now to secure the purchase or wait until next week to take advantage of the 10% markdown but at the risk of not finding the right size?

In one study the variables used on this scenario were the different degrees of scarcity: high (only 2 pairs available) or low (10 pairs available) and future discounts (low: 10%, moderate: 25% and high: 50%).[7]

The readiness to buy immediately grew by 34% where there was a high degree of scarcity. In addition, a low future discount increased the probability of an immediate purchase at full price. The same was true in a situation of scarcity with high markdowns.

A combination of scarcity with future discounts thus favors future sales.

Experiments like these show how susceptible human beings are.

The fact that Amazon's website indicates, for example, that there are "Only 2 immediately available" of several bestsellers is certainly nothing to do with the service provided, which ought to be the only discriminating factor for an e-commerce business.

Several studies confirm that customers judge products in a casual or arbitrary manner: at times based on a visual stimulus such as a price with a cross through it alongside a lower price or, as in the above case, when the quantity available is reduced in order to generate impulse buying. People who tend to behave wisely in everyday life may buy quantities of a product that is scarce or on discount without checking whether the offer really is a bargain.

7. Using a sense of winning to break down purchasing barriers

Customers do not usually realize how far they are manipulated by salesperson's tricks. This is because of a series of unconscious processes that exist in the human brain.

Academics like the Nobel prizewinner for economy Daniel Kahneman confirm: the emotional reaction to loss – payment of a price – can be far stronger than the reaction to winning, for example the joy of possessing a new car.

This emotional asymmetry lies at the heart of the theory, which explains certain "pricing structures" that would otherwise seem absurd.

The popular *cash back* for the purchase of an automobile in the United States is one of these examples: if you buy a car for $30,000, at the same time you receive an incentive of $2,000.

According to this theory, the buyer is suffering a loss – he is spending $30,000 on the automobile – yet he will experience the victory of realizing the incentive even more strongly than the ownership of a new car.

If the customer, who generally pays by bank transfer or check, receives $2,000 cashback – that is, the cash is put directly into his hands at the moment of payment – the sense of victory from this incentive even seems to be superior to the price paid.

On reflection this is banal and might happen to anyone: they give you 2 whilst you pay 30, yet it feels like they've given you 2 (not that you've paid 28). What a weird sort of animal humankind!

Although this theory may seem bizarre, it is in fact more than confirmed by the facts. Otherwise, how would it be possible to find so many high prices on the lists, which no-one actually pays?

It would be more rational to offer a product at $75 instead of pricing it at $100 and then allowing a discount of $25. And yet in many customers' minds, obtaining a discount generates a sense of winning. And this is why a number of companies substantially increase the prices on their lists, so as to then offer constant markdowns; in other words, they sell their customers the bargain, or discount.

It is the sense of "winning" that guides the purchase, nothing else: it is the (lowering of the) price.

8. Optimizing the relative discount versus the absolute discount

If I sell a product and decide to apply a discount from $85 to 70, how best can I indicate the markdown: as a price reduction of $15 or as 18%?

Numerous studies indicate that customers react differently when the same markdown in absolute terms is shown as referring to a different price.[8]

This is demonstrated by the following experiment[9]: a customer is buying a jacket for $125, and a calculator for $15.

The seller of the calculator immediately informs the customer that the latest model can be found on offer at just $10 at a sales point belonging to the same chain, a 20-minute drive away. With a 33% markdown – that is, $5/$15 – as many as 68% of customers are ready to get into their cars and drive off to obtain the discount.

But when the seller of the jacket informs the customer, in the same way, that the same, identical jacket can be found at only $120 in another shop belonging to the same chain as the one they are in now, 20 minutes away by automobile, only 29% of customers are ready to get into the car for a markdown of 4%, that is, $5/$125.

Therefore.

It is preferable to express discounts on products sold at lowish prices in relative terms, that is, by means of a percentage, whilst it has been demonstrated that customers prefer discounts in absolute terms for products sold at prices perceived as high.

If the price on the list is over $100, it is therefore best to apply an absolute discount rather than a percentage.

Discounts should be used cautiously for high-quality products. Discounts on high-quality products where buyers do not expect to find price-based competition (paradoxically) turn the customer's attention precisely to the price as the criterion for purchase.

When, on the other hand, customers attribute more value to price as their criterion, products with lower prices suddenly become acceptable alternatives, despite their lower quality.

By extension, discounts on high-quality products have driven customers towards lower-quality offers, whilst the contrary has never occurred.[10]

9. Impacting price perception by means of visual design

"It has to look pretty, too" they say in the art world, and this applies to pricing, too.

Elements such as the size of the letters, color, and special offers all influence price perception.

A typical way of managing promotions is to make the lower price catch the eye immediately by using larger letters compared to the starting price, for example. But it is not the most advisable approach. This is because, from a psychological point of view, it is easier to associate lower prices to numbers in smaller print, compared to higher prices in bigger print.[11] Some specific analyses indicate that when the lower price is presented in smaller letters than the normal price, the perception that remains is that of a bargain, which has the positive secondary effect of increasing the customer's intention to purchase.[12]

Color, too, can attract buyers. For example, it has been demonstrated that if you're male and the prices are shown in red, there is a high probability you will associate this with a better bargain.

In general, people follow two paths when processing information and taking decisions: a systematic approach or a heuristic one.[13] In the systematic approach, decisions are based on careful evaluation and taken in fully awareness. When "rule of thumb" applies and plausible hypotheses are followed, we find ourselves in the context of heuristic decisions. The former requires a greater cognitive effort, whilst the latter is a mental shortcut.[14]

Lastly, choosing the best path to follow when making a decision depends on how deeply involved we are. Some studies show, for example, that people are less involved when they process information coming from advertising using typically heuristic indications.[15]

In the study examined here, participants were presented with advertisements for toasters and microwave ovens. In terms of semantic trace, the text was written in black whilst prices were highlighted in red. The result was the following: to sum up drastically, if the prices were shown in red, men perceived a better bargain.

In this specific study on the male target greater positive emotions were in any case associated with red prices, compared to the same prices listed in black. Nevertheless, as soon as there was a shift to more involving decisions, the effect of the red disappeared.

The female public, on the other hand, proved to be immune to any influence of color in prices: no difference in perception was recorded when the color varied in similar tests.

And if the whole truth were told. . . when the sales are on, how many people ask themselves whether their purchases really are bargains?

It has been shown that the mere indication of a cut price, for example by means of a label with the words "on sale," increases sales.[16]

Therefore, offering a mix of promotions with negligible discounts associated with the sale of highly discounted products increases profitability.[17]

Summary

Behavioral pricing can make a difference: alongside rational decisions, there exist numerous, irrational factors that determine purchasing behavior. These apply both in a B2C context and, to the same extent, in B2B.

In behavioral pricing practice we come across some basic rules that help companies monetize the value delivered to customers in a targeted manner, such as the following:

1. Contextualizing the value with a "price anchor"

2. Breaking down resistance to buying using the printer/ cartridge model

3. Taking advantage of maximum readiness to pay by using "threshold pricing"

4. Facilitating customers' choices by the compromise effect

5. Using pricing as an indication of quality

6. Creating scarcity to encourage sales

7. Using the sensation of winning to break down barriers to buying

8. Optimizing relative discount vs absolute discount

9. Impacting price perception by means of visual design

The nine tactics suggested offer some practical inspirations, but obviously before applying them, it is always as well to check and hope you don't come across the devilish Matt-and-Harry act!

CHAPTER 6
DYNAMIC PRICING

"We want the offer to always be available and price is used (basically) to reduce or increase offer . . . it's a classic in economy"

Travis Kalanick, CEO and co-founder of Uber[1]

Case History

The pandemic. The wildfires. The war in Ukraine.

Perhaps we don't want to say or write anything any more.

Perhaps it's true that the planet as we know it no longer exists (or will exist).

Yet there's a part of us, and not even such a small one, that doesn't want to give in to defeat.

Perhaps.

When the coronavirus is over – because everything comes to an end sooner or later – when the frontiers are open again, unencumbered by those who want to build walls instead of bridges, when we can travel again, wander the world, to a concert for harpsichord by Johann Sebastian Bach or else the Milan–Tel Aviv route without thinking too much about it, it will mean that the meaning we have in mind for the word "world" will be reality again.

A world of mobility and quick exchanges where it's possible to *act*.

Another idea of travel, a new idea of the future and the enjoyment of places, whether for work or on holiday.

And when we get back there – tomorrow, but basically humankind is always the same – we shall (also) find ourselves up against the old "questions," as Shakespeare would say; up to yesterday, for example, booking a summer holiday just before your departure, could be, if not a nightmare, at least a very expensive experience.

For a flight to the most popular summer destinations the price could/ can more than treble in the peak season, since the airlines raise their prices as the warm season, fun, leisure time and so on draw nearer, because we all know what everyone's after in summer. All that can be expressed aloud that is.

And so.

If demand grows for seasonal motives, airlines, hotels, and tourist operators exploit customers' greater willingness to pay or, as Robert Crandall, former CEO of American Airlines, would say: "If I have 2,000 customers on a route and 400 prices, I'm obviously short of 1,600 prices."[2] What Crandall means is that the best thing in this case would be to have 2,000 prices – thus changing the concept of fixed pricing – or targeted, personalized pricing, able to correspond to every user's specific demands: applying a high price in the case of the business user who makes a move the day before, for example, and in the case of the user who books a year ahead, offering them the best price possible.

A sporadic phenomenon in the analogical world, the "adaptive" pricing system got off the ground thanks to the advent of Internet and the new technologies; those who buy on Amazon or eBay have to reckon with a daily or even hourly shift in price.

On Amazon prices change on average every 10 minutes, or 144 times a day.[3] In the space of just a few hours, the same price can vary up to 240%.[4] And so, for example, Amazon puts up prices in the evening or at weekends, when people have more time to devote to buying and the demand increases.[5]

This is the essence of dynamic pricing: the selling price of a product adapts to the contingencies of the market: if the demand for a camera increases, so does the price, since there will be a greater demand and the customer will therefore be more willing to spend money in order to get hold of one of the few remaining items; if, instead, the demand falls, so will the price, precisely in order to encourage the demand for that product, which up until then had not made any headway in the customer's needs/desires (in both cases boosting Amazon's profits by 25%).[6]

The cost of fuel at a gas station also changes throughout the day, sometimes even while you're filling up; this is another example of dynamic pricing, also known as "dynamic price management." Behind all this lurks a concept as old as trade itself: dynamic or flexible and versatile/adaptable prices according to the market situation contribute to controlling sales of products and services.

What will change increasingly in the future, thanks to the new technologies, are the frequency of the changes and the quantity of products offered.

One of the most famous cases of how an "innovative" company can succeed in upturning a whole sector using the right form of monetization – dynamic pricing in this case – is *Uber*.

Before becoming one of the leading players in the mobility sector, the management of Uber discovered a fundamental truth: the flexibility of the "law" governing any sort of move from A to B, is particularly high. But – and the dynamic intuition is here – this is not only true on the side of demand, that is, from the user's point of view but, Uber realized, it is also true on the side of the offer, that is, of the drivers.

The problem to solve in the management of peak demand after a sports event, a music festival, or on a Saturday evening, when the demand for transfers was (but we have said we're getting back to live concerts again, and so. . .) *is* so high that there is a lack of vehicles, the problem is to find more drivers!

The Uber managers' solution from this point of view was really "smart": they created a new model of monetization – *surge pricing*.

What does it comprise? Let's look at the main aspects.

It is a form of dynamic pricing that sends prices shooting up where there is an excess of demand, thanks to the latter being monitored in real time. The price increase thus has a dual effect which is fully intentional.[7] On the one hand, it attracts a larger number of drivers who, at lower prices, would not have been on hand. On the other, it brings down demand, balancing it with the offer, so that prices can settle again.

In this way Uber created a platform able to conciliate demand and offer; there is no obligation imposed as to when drivers offer their services, neither are slots scheduled when they must be on hand, let alone shifts.

This is all quite normal in the case of a taxi-drivers' cooperative; in the case of Uber, however, everything is regulated by the pricing:

by means of the technological platform millions of rides a day are managed without any instructions being given to the drivers, who decide independently how to make their moves on the basis of dynamic pricing.

In other words, the offer is not controlled directly by Uber but by the many independent drivers.

In the end it is Uber's customers who decide whether or not to accept the price offered which, in extreme cases, may even be nine times higher than normal.

The alternative, in any case, would have been to maintain a lower price that would not correspond to an adequate offer (thus provoking the anger of customers willing to pay more to have a reliable service with drivers available).

All the same, *surge pricing* has met with a lot of controversy.

If too many customers reject higher prices n times compared to normal, Uber quickly intervenes to re-set the price.

Briefly, the primary objective of this dynamic pricing is to guarantee sufficient availability of drivers, even at times of peak demand: most of the higher price ends up in the pockets of the drivers. This is the only way an offer adequate to demand can be obtained 365 days a year, even when other means of transport are scarce – because of last-minute strikes or atmospheric reasons, such as hail or snow. Compared to the much discussed *surge pricing*, the other side of the coin in dynamic pricing is the downward trend in prices.

If the drivers are sensitive to price, there is no doubt passengers are too.

THE PRICING MODEL REVOLUTION

New York.

Coppersmith, a seagull shakes water off its feathers.

A barman makes coffee in a café.

Pages from a notebook flutter in the wind.

The subway swallows and spits out managers, young people are crossing on the zebras.

Cut.

One of Uber's inventors is driving through the night.

The images flash past again: customers at a restaurant, eating *tartare* at candlelit tables, the disc jockeys are playing background music that we can't quite hear but easily imagine.

Friday, Saturday, Sunday, the weekend, time rushes past below our lives.

The man at the driving wheel tells us of the "simplicity" of moving from the famous points A to B.

New York again.

Girls raise their hands, the taxis don't stop.

Men in suits and ties, black car in attendance, explain that the problem of transport exists in every city in the world.

Skyscrapers, Brooklyn Bridge.

Uber's initial slogan, "Everyone's private driver" was precisely aimed. Efficiency, comfort, convenience. And the drivers' downtime, when they're without passengers, waiting for the next ride. Whilst there are many people who might need them. How can these two opposite and complementary needs be reconciled?

The idea the entire company was based on was precisely that of offering cheap, safe, ever-available rides through an extraordinary purchasing experience: by using the app, you know exactly when the driver will be coming to pick you up, you can see where and how the driver is proceeding and in addition you know in advance exactly how much you'll be paying and the transaction can easily and quickly be made by credit card.

By using dynamic pricing in the opposite direction to surge pricing – in practice by offering a penetration price – Uber succeeded in raising the average running time of its vehicles, also generating a demand from some segments of passengers who, at a higher price, would have preferred to own or hire a car or use public transport, rather than move around *pedibus calcantibus*.

The company's management goes even further: why not focus on increasing the use of the drivers' vehicles to 100%, thus attaining what the CEO, Travis Kalanick, calls the "perpetual ride,"[8] that is, having at least one passenger for the entire shift.

This means, on the one hand, optimizing rides so that one passenger starts their ride as soon as the previous one reaches their destination. On the other hand, it even means encouraging customers to sell their own cars, owned but unused (for an average 95% of the time), in order to replace them and establish Uber as the prime means of transport.

A first step in this direction is *UberPool*, an offer addressing those who move in the same direction or plan similar routes.[9]

An interesting offer regards all those who have to reach the outlying Heathrow Airport from the center of London or, even, in another part of the world, tourists who want to get from the Vatican City to the ancient Roman remains in Ostia Antica, which are just as far from the center.

In both cases, the vehicles are better filled and their use rate increases. In addition, and most importantly, the price can be reduced, making it proportionally more economic.

And so Uber becomes not only more advantageous than your own or a hired car, but also competes with public transport.

From its timid début on the streets of San Francisco, Uber has grown to establish a presence in 50 countries, becoming the main new player in transport, rated well above giants like General Motors,[10] and the critical factor in this success has been dynamic pricing, examined in greater detail in the next section.

Analysis of Context

Origins and development

Dynamic pricing has and will increasingly become of fundamental importance to companies, and in the most widely varying sectors. Even big distribution, which until a short while ago displayed fixed prices on the shelves, is introducing electronic displays that make it possible to change the prices at the retail point several times a day.

MediaWorld, one of Europe's main distribution chains not dealing in food, has thus shifted to dynamic pricing management not only in its online channels but also in those offline (their retail points), always indicating the best price to their customers.

Dynamic pricing management is an ancient phenomenon, whilst it is fixed prices that are relatively new: they did not appear until "recently" and before 1870 it was quite normal not to display prices but to vary them dynamically, so that every price was individually negotiated.

The Quakers were amongst the first to consider it immoral for different customers to pay different prices, thus starting to put actual labels onto articles and doing away with bargaining. And so big stores like Wanamaker's in Philadelphia and Macy's in New York introduced fixed prices. This made it possible to reduce the time it took to train sales staff, who were no longer obliged to know the prices of goods, improve the art of bargaining and the market and enabled more customers to be served and sales to become more efficient. Even the big stores kept goods behind a counter and asked sales staff to go and fetch articles for a customer for each individual sale. This implied downtime that we could no longer afford today.

The Piggly Wiggly chain of food stores was the first to go self-service in 1916, further confirming the need for labels. The price labels soon became a common way of displaying goods on sale in Western retailers and bargaining was gradually limited to sellers of second-hand goods.

And so dynamic price management slowly went to sleep for over 100 years.

The awakening of dynamic pricing came about in the United States of America in the 1980s with the liberalization of prices on passenger airlines, which had up to then been regulated by the government.

From then onwards, airlines reverted to managing the most important profit driver: pricing.

Liberalization encouraged the advent of low-cost companies, contributing to the growth of the whole sector and attracting price-sensitive customers, who would otherwise have used other means of transport such as trains or automobiles to move about.

It is therefore evident that in this sector, elasticity of demand according to price is a decisive factor, in particular in the downmarket segment, and it is a factor that makes ample growth margins possible.

Proof of this is People Express Airlines, founded in 1981, which in 1984 reached the billion-dollar mark by offering prices that were 50% lower than those of traditional airlines, also obtaining a new record for the times in terms of profit: $60 million, an absolute record for this company, whose fate we shall return to later on.

Meanwhile, traditional airlines such as American Airlines in the same period – or rather, from that moment onwards – lost a considerable number of (price-sensitive) passengers who "migrated" towards low-cost companies. The need to put into practice new business strategies to regain lost customers became evident.

Given the significantly limited cost of the low-cost carriers, traditional companies knew they were up against a tragic dilemma: whilst at that price level companies like PeopleExpress worked with positive profit margins, for them a rush towards lower prices would have meant heavy losses. To implement this new strategy, American Airlines had to face two challenges: establish the number of seats to allocate to economy compared to business class and avoid the cannibalization effect, that is, the use of low-cost seats by travelers willing to pay higher prices.

The introduction of "super saver" tariffs and formulas characterized by a few restrictions (such as purchase 30 days before traveling, a stay of at least 7 days and no reimbursement) combined with a carefully fixed quota of low-cost tickets per flight, made it possible to propose a valid alternative to PeopleExpress and, at the same time, not to compromise the (far more) profitable business segment.

It was American Airlines' then Head of Marketing, Robert Crandall, later promoted to CEO of the company, who first identified the way out of this challenge. Crandall realized that his company was selling seats at zero margin costs, because most of the costs of a given flight are fixed: for example, the pilots' and air stewards' salaries, amortization, fuel and so on.

This new logic allowed American Airlines to compete with low-cost airlines by, for example, offering some of the seats not occupied by business travelers at their prices when unoccupied by business travelers, characterized by tickets purchased at short notice and higher willingness to pay.

With the development of DINAMO – acronym of "Dynamic Inventory Allocation and Maintenance Optimizer" – American Airlines solved the problems of allocating capacity to the different classes of offer per route and flight. Moreover, this system, which came into operation in 1985, represents one of the first real dynamic pricing systems in the business field, also known as *revenue* or *yield management*.

In fact, DINAMO enabled aggressive competence, combined with the possibility of rapidly modifying offers on specific flights.

On all the routes served by *AA*-American Airlines and low cost airlines an authentic price war broke out. In particular, the blow DINAMO delivered to PeopleExpress proved deadly: in 1986, only a year after the introduction of the *AA* Optimizer, PeopleExpress went bankrupt and was bought out by Continental Airlines.

PeopleExpress's CEO at the time, Doland Burr, explained the causes of the bankruptcy as follows:

> *From 1981 to 1985 we were a dynamic and profitable business, which then went on to lose 50 million dollars a month. It was still the same company. What changed was American Airlines' ability to implement revenue management on each of our markets. We generated profits up until American Airlines' introduction of the "Super Saver" tariff, which abruptly marked the end of our run, because it gave them the possibility of offering lower prices than ours at their discretion. We can't deny that PeopleExpress went bust [...] Nonetheless we did make a series of good choices. What we got wrong was that we neglected dynamic pricing. If I could start over, my first priority would be to have the best technological support system possible. For me, this is what is decisive in producing revenue for an airline: more than its service, more than its aircraft and even than its routes.*[11]

Main forms of dynamic pricing

Three main forms of dynamic pricing can be distinguished: (1) temporal; (2) customer-based; and (3) based on sales channel.

1. *Temporal dynamic pricing* is considered a "take it or leave it" sort of pricing, where the seller changes the price as time passes, on the basis of factors such as sales trends, the evolution of demand and the availability of the products in demand. Here, dynamic pricing coincides with the revenue management of airlines: the prices vary so as to make demand coincide with offer. The same thing can be found in the energy sector or in

fuel distribution. Dynamic pricing based on time is described mainly in terms of frequency and range.[12] *Frequency* refers to the number (how many times) of price changes, which may be considerable. Amazon changes all its price points 2.5 million times a day, meaning that an average product's cost will change about every 10 minutes.[13] Changes can also be more frequent: the price of a mobile phone was changed 297 times in three days on Amazon.[14] *Range* describes the amount of individual price changes.

2. The second form is *dynamic, customer-based pricing*, which goes under several names: "personalized" pricing,[15] "behavior-based" pricing,[16] "targeted dynamic" pricing,[17] or even targeted promotions.[18] The basic idea of this tariff is to exploit, to the utmost, customers' different degrees of willingness to pay: not knowing exactly how much their customers are ready to pay, companies resort to items of information considered to be indicators correlated to willingness to pay. For example, they might use demographic data or data on surfing behavior, information on past transactions or again analyze data on the *customer journey*, for example if customers have come to the website through online comparisons of prices.[19]

This is why Uber, Lyft, or Airbnb offer incentives to registration and discounts for new customers; personalization is based on the status of the customer, new or already acquired.[20] The basic prerequisite is that customers can be identified. Whilst in the offline world it is often necessary for customers to be members of a loyalty program, in the online world other data can be used. What is more, two central forms of customer-based dynamic pricing can be distinguished. On the one hand, companies can offer different basic prices; on the other, they can use personalized coupons.[21] Using these, the base-prices remain constant on the web. Nonetheless, some selected consumers[22] or groups of customers[23] receive discount coupons that make it possible to calibrate prices to their willingness to pay.

3. Another form of dynamic pricing presents itself for multi-channel companies, that is, those with *both offline and online channels*. These companies find themselves facing a "multi-channel pricing dilemma": through their online channels, they are particularly exposed to pressure from the prices of pure e-commerce retailers. The cost structure itself of offline channels – costs of rents and personnel – makes their life difficult, as it produces internal competition between online prices that are generally lower than those offline.

Therefore, the idea of dynamic, channel-based pricing is to differentiate between offline and online prices. Prices that vary according to channel may seem justified by the different functions offered, for example: whilst online channels allow saving on travel costs and may offer a wider assortment, offline channels are distinguished by the possibility of seeing and trying out products. Other examples of company practice serve to illustrate dynamic, channel-based pricing, as in the case of the consumer electronics company *Conrad*, which announced one price in shops and a lower one online on its website[24]; in the same way, Lufthansa or Northwest Airlines Corp sell air tickets online, over the phone, and offline. Nevertheless, frequent changes in price are almost exclusively made on their online channel[25]; those who buy offline enjoy the reassurance of a service corresponding to their expectations, online instead implies more careful monitoring by the customer but also the opportunity and the sensation of managing to "win" a lower price.

Expansion and impact

Dynamic price management is on the increase, particularly in online business. As time passes, however, it has also taken root in the sector of air travel, holidays, and hotels. Factors such as use rate, season, schedules, and comparison to competitors influence prices. Even in traditional shops, the usual price tags are increasingly replaced by digital displays on the shelves, allowing simpler management and high automation.

Since its introduction, dynamic pricing has been used throughout the air travel industry and made a substantial contribution to profits. Implementing dynamic pricing can improve revenues and profits by up to 8% and 25%, respectively.[26]

Alongside the B2C sector, it can be seen that an increasing number of B2B companies are adopting dynamic pricing. For this, please consult the book *Revenue Management in Manufacturing*.[27]

Success factors

It doesn't always make sense to make use of dynamic pricing and, in fact, if the premises are lacking, it may even be harmful to resort to this tool.

The adoption of dynamic pricing is quite a journey for any organization: perseverance is needed as well as readiness to overcome all the obstacles that will inevitably come up along the way. But if it is well prepared, incredible acceleration will be experienced and from this point onwards the objectives of implementation can successfully be attained.

In particular, on the basis of the numerous projects studied over the past few years, we can state that four success factors emerge: the first two regard the solution and the second two the integration of this solution into the company.

1. *Data and technologies*: the first premise for setting up a dynamic pricing model is the availability of data and information. In some sectors a high level of granularity is required and this is

the case in B2C, even with regard to individual customers or transactions; in others, such as for example that of an electrical materials wholesaler selling to electricians, it may be enough to use data relating to a segment with a market channel. What counts in both cases is the availability of an IT architecture ready to support the new mode of monetizing the value provided to the customer.

2. *Price logic and pricing tools*: the logic, that is, the way in which prices are set in companies today, is the starting point for evolving towards dynamic pricing. This may differ a lot from company to company. In the B2C context, a retailer can vary prices according to customer loyalty or by the day or by the time slot of the purchase, or by shopping basket. In the B2B context we may be dealing with telephone operators who base the price on parameters such as the pricing of their competitors, the duration of calls or, for example, on data use schemes. What will be useful to every company is the development of a dynamic pricing tool that comprehends the application of specific logic and equations able to develop dynamic prices.

3. *Process and government of dynamic pricing*: "A fool with a tool is still a fool," said Grady Booch, the famous IT engineer who for years, many of which spent with IBM, has been working on collaborative dynamics in IT environments (he was the speaker at the 2007 Turing Lecture entitled: *The promise, limits and beauty of software*): although "centered" both on pricing logic and a powerful analytical tool, the success of dynamic pricing is determined by the way people in the company use these tools: it's the teamwork that makes a difference to the game! The pricing and marketing functions provide the impulses for calibrating and constantly improving the approach; sales bring their knowledge of customers and markets; the IT function, with its technical competence, helps to tweak the power of the tool; and the *control and finance* function monitors the results,

which will then be submitted to the heads of the company to obtain future guidelines.

4. *Habilitation and competences of the team*: last, but no less important, come the habilitation and competences of the teams. The most successful transformations in pricing are those where the heads of the company, often the CEO, call on the whole organization to support the new dynamic model. Management's role is essential to ensure cohesion in the multidisciplinary teams.

Lastly, special habilitation work and action for instilling a "culture of openness, trial and learning," facilitates things a lot. Often, companies that have had success here have started out from single applications – the so-called *use cases* – later extending the perimeter after having tested and gathered experiences that have contributed to enriching the team's competences.

Expedients for avoiding non-optimal, or excessive prices

Technological advance over the past few years has provided solid bases for raising dynamic pricing to new levels.

Thanks to the spread of information such as data on individual preferences, buying behavior, demographic characteristics, prices of the competition, modes of payment and so on, customer profiles can be created on the basis of demographic data and buying history and all the marketing tools can be duly calibrated.

Nonetheless, this must not remain a lever that is managed by computers alone: good sense is required in order to avoid offering prices that are less than optimal or, in certain cases, quite inadequate.

Let's turn once again to the company considered to be a benchmark in dynamic price management: Amazon. Although it is capable of changing 2.5 million prices a day, maximizing profits and becoming one of the most valuable companies in the world, it does not always offer optimal prices.

We might wonder, for example, why Amazon sells a Samsung TV first at \$296.99 and then at \$293.07, if psychological pricing – which we spoke about in Chapter 5 – teaches us that these two prices are perceived the same way by the customer as \$299.99, thus needlessly wasting a profit margin.

Another striking case is that of the amazing price demanded for a book on flies[28]: dynamic pricing inflated the price of *The Making of a Fly* to almost \$24 million, not including delivery, of course (!), as shown in Figure 6.1.

How was this astronomical price arrived at? Easily explained.

By means of dynamic pricing, which compared the price of two similar books but, whilst one equation continued to set the price of the first book at 1.27059 times the price of the second, the other equation automatically set its price 0.9983 times the price of the other. And in this way, in a crazy dance of hysterical rises in price, the prices of the two books grew together until they had reached the millions, with the second book nonetheless conserving a slightly lower price than the first.

An authentic marketing error which leads to a fitting conclusion: Dynamic pricing management has a limit – it's as good as the equations that regulate it.

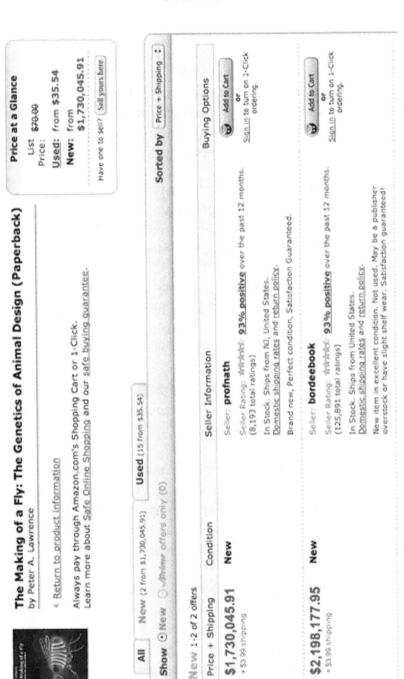

Figure 6.1: Worst practices of dynamic pricing: price explosion for a book on Amazon

Summary

History repeats itself, said Karl Marx and "History Repeating" sang the Propellerheads, too , in the song (featuring Shirley Bassey) of the same name: dynamic pricing was the norm in the past, before the advent of fixed price labels in 1870, later re-emerging in the sector of air travel and spreading to the widest range of industries.

The recent renaissance in dynamic pricing is not, however, an event guided primarily by negotiating capacity and taking place between the seller and the customer with an unpredictable outcome, but rather an event fueled by technological advances.

The proliferation of data makes it possible to segment customers according to demographic clusters, buying behavior, or prices of the competition, at the same time refining marketing tools – from targeting to individual prices – which vary over time.

This goes both for the B2C context and for B2B: it can be applied in both but it doesn't always make sense to introduce it in every context and it should therefore not be "forced."

In practice we distinguish three main forms of dynamic pricing: temporal, customer-based, and based on sales channel. Dynamic pricing finds its application in the widest range of sectors, involving both service companies (e.g. tourism or car hire) and manufacturing companies (e.g. producers of steel or chemical products).

Once introduced, it makes a considerable impact: dynamic pricing can improve revenues up to 8% and profits up to 25%.

In order to have dynamic price management, experience tells us that there are essentially four premises: the first two regard the solution chosen, that is, *data and technologies, price logic and pricing tools,* whilst the latter are to do with integration of the solution into the company, that is, *process and government of dynamic pricing and training and competences of the team.*

In general, the usual warning still holds: dynamic pricing is a tool and must be managed as such! Machines and equations alone are not sufficient: they serve human beings, who must make sure that the results make sense (see the "book on flies" affair) and thus generate optimal prices. History – as well as repeating itself – will take care of the rest, its own course and that of our business.

CHAPTER 7
ARTIFICIAL-INTELLIGENCE BASED PRICING

"An android life is dream."

Philip K. Dick

Case History

Mac Harman scented the business opportunity just after graduating from the Stanford Graduate School of Business: he had observed that his in-laws' fake Christmas tree didn't look much like a tree and that was when the idea came to him.

He set off for China, where he met a tree producer and designed 16 models based on a variety of Christmas trees that were the shape of a real "Christmas tree," like a spruce, for instance.

In October 2006, he had more than 5,000 trees sent to the United States, where he opened a pop-up in a mall in Stanford. Business went very well.

After creating a website, in the space of one month he had, in fact, already invoiced $3 million. Since then the good "Mac" (and perhaps it really is the name that creates the destiny) has broadened his selection of trees, some of which are sold at over $2,000, and has added decorations, stars, garlands, and other products.

Balsam Brands – this is the name of his company – was success-ful, despite selling seasonal products at prices far higher than their direct competitors.

Since then the start-up founded by Harman has grown even further, and today faces new challenges with regard to pricing.

The Californian retailer of luxury fake Christmas trees and seasonal decorations – whose brands include the popular *Balsam Hill* – earns over $200 million, 80% of sales taking place during the last three months of the year. This is a managerial and financial "asymmetry" of crucial importance to the company's business activity.

To manage the seasonal sales and at the same time protect the profit margin and make revenue grow, Balsam Brands has therefore decided to introduce artificial intelligence (AI) into their pricing.

The algorithm is self-training, generates recommendations, opti-mizes prices on the basis of demand, and thus overcomes all the challenges that previously existed, which included a particularly intense repricing process, as well as the lack of an overall tool for managing prices on the basis of market trends.

Using a personalized ERP platform, Balsam Brands has automated weekly repricing, adjusting prices according to the business plan and taking decisions based on the data processed.

To make their best offer to their customers, Balsam Brands takes into consideration a number of price factors and more, such as web analyses and market trends, the latest sales figures, price scales, intelligent business constraints and the rules of price rounding

During the 2020–21 trading season, the AI algorithm generated 24,000 recommendations on optimizing prices for the retailer.

They were based on its historical transactional data, on trading constraints, on the architecture of total pricing, the availability of the inventory, and other essential information. As a result, Balsam Brands cut its repricing time by 50% and reached its pre-established business objectives, generating over 3.5% extra revenue and an over 3% extra profit margin: "While our business was growing, it was important for us to base our pricing decisions on market trends, website analysis and other crucial data that's difficult for a pricing manager to take into account simultaneously," stressed Joyce Lin, senior e-commerce business manager of Balsam Brands. "Intelligent algorithms have made our pricing management efficient, helping the team save 50% of the time spent on routine tasks. AI is revolutionizing traditional pricing strategies and processes, so we can't wait to extend this technology to other regions."

The same direction was taken by Orsay, a textile company present in 34 countries with 740 sales points and 5,100 employees[1]: "Today we no longer have to depend on manual analyses or conjecture. AI-based pricing has automated our most critical pricing decisions. The algorithm makes a recommendation and we simply apply it," this just about sums up the thinking of Orsay's Chief Innovation Officer.[2]

As a vertically organized, fast fashion retailer, Orsay manages the entire supply chain, from design to production, to sales.

The company offers a detailed range of products: a vast selection of trendy and classical styles.

As fashion trends are continually changing, Orsay has to manage the pricing of their products for their entire life cycle: maximizing profits but ensuring that the clothes are sold before they become obsolete.

Briefly, Orsay's objectives were:

- to increase income and profit margins using fewer markdowns;
- to reduce inventory costs by getting rid of stock more efficiently;
- to improve staff productivity;
- to increase customer satisfaction by making their product expectations coincide with pricing.

This has determined a situation in which: "In the first year of using AI alone, our markdowns were reduced. The percentage of stock per markdown has improved by 3–40%, from an initial range of 40–50%. This means a markdown percentage lower than 10%, which raises our profit margins. Today we are able to sell our products when there's a demand for them. We apply discounts less frequently. In the past we might have applied up to three or four markdowns per article. Obviously each of these eroded our profit margins. Today we apply a maximum of two or three discounts per article."

In Orsay, AI is also able to reduce waste and obtain the best prices constantly, guided by data in each phase of the product's life cycle.

Considering present and historical data, the algorithm thus determines the right level of price elasticity for each item of Orsay's clothing.

In addition, the solution takes into account complex factors such as competitors' pricing, the effects of replacement and cannibalization, whilst automatically taking the most profitable pricing decisions for Orsay.

The international fashion company Bonprix,[3] present in over 30 countries, with 35 million customers, has also shifted to AI-based pricing. Folke Thomas, responsible for introducing AI in Bonprix, in one of his interviews recalled the start of the new pricing approach: "Since then, variations in prices can be implemented by means of the algorithm from one day to the next, for example if there are short-term needs in managing profits or stock." They no longer have to be discussed and aligned, involving numerous functions within the company, and then implemented manually.

The same goes for Orsay: in the past the average category manager spent up to 80% of their time dealing with markdowns. Machines have automated this aspect and now the same item only "weighs" for about 20%. In the time saved, more attention can be devoted to strategic content.

Analysis of Context

AI is a machine's ability to demonstrate human capacities such as reasoning, learning, creativity, or planning. In other words, IT systems able to carry out tasks that would normally demand human intelligence, such as visual perception, voice recognition, decision-taking processes, and translation from other languages.

In price management AI comes from algorithms able to identify optimal prices or markdowns by analyzing the effects of past business policies, for example, or considering a range of further information that makes it possible to learn and calibrate pricing and adequate discounting.

AI-based pricing therefore means using AI methods like machine learning and deep learning to imitate human behavior and take autonomous pricing decisions that improve constantly, using statistical methods and advanced algorithms (see Figure 7.1[4]).

The technology of machine learning is contributing to great changes in the field of best prices, since it is able to manage pricing at a far

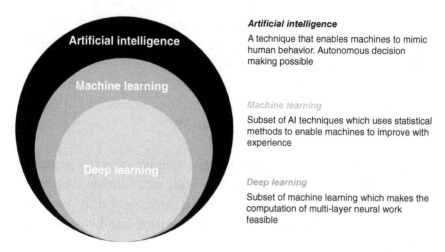

Artificial intelligence

A technique that enables machines to mimic human behavior. Autonomous decision making possible

Machine learning

Subset of AI techniques which uses statistical methods to enable machines to improve with experience

Deep learning

Subset of machine learning which makes the computation of multi-layer neural work feasible

Figure 7.1: What is AI – a definition and separation from machine learning and deep learning

quicker rate with much greater effectiveness. For example, algorithms based on machine learning are able to analyze huge amounts of data simultaneously and take into consideration more variables than would be possible without AI.

In the past, pricing managers had to determine the rules of price management manually. Instead, machine-learning models use algorithms that constantly learn automatically from their results. Companies are therefore able to use self-learning models to set prices or adapt them over time; in addition, they can do this quite independently, very precisely and with a fraction of the effort.

AI-based pricing tools are not only designed to learn, but to improve with time in finding the peaks of best prices, since they are capable of distinguishing a price that is calibrated between "too economic" and "too expensive." Moreover, AI-based pricing tools can consider both critical internal data and external data that influence their algorithms. This, combined with the fact that they can elaborate huge and diverse series of data compared to the old technologies, allows them to be extremely accurate as to the prices they establish in relation to influential data.

The factors evaluated by these algorithms include:

- historical sales and transactional data;
- seasonal changes;
- weather conditions;
- raw-material price indexes;
- geographical data;
- events;
- inventory levels;
- product features;
- prices and promotions offered by competitors;
- customer relationship data;
- marketing campaigns;
- reviews and articles.

Using this data, the price applications based on AI can calculate elasticity of price, measuring how demand will fluctuate as a result of changes in conditions. The software regulates prices consequently.

These tools can also determine the products for which demand is stable enough – making them suitable for optimizing profit margins – or those that play a critical role in overall sales and must thus be carefully regulated.

Phases for Developing AI-Based Pricing

How exactly does the creation of an AI algorithm for determining best prices work?

Although it might seem obscure to some, in fact the steps towards setting an algorithm for optimizing prices based on machine learning are straightforward. The process works as follows:

1. Data collection and cleansing

To develop an automatic machine-learning model, several types of data are needed.

In the context of best prices, the database might look something like this:

- *Transactional data*: list of products sold at different prices to different customers, types of discount allowed on invoice or not, premiums and bonuses;

- *Description of products*: data on each product catalogued (category, brand, size, color, etc.);

- *Data on cost*: cost of supply, cost of delivery, cost of returns, cost of marketing;

- *Data on competition*: competitors' prices for comparable products either entered manually or derived automatically e.g. via webscraping;

- *Inventory and delivery data*: data on inventory levels, product availability, price history.

Not all this information is necessary, neither will it be available for every sector or business. For example, many retailers do not have a "clean" series of historical data on prices. Nonetheless pricing based on machine-learning is able to extract the maximum of intuitions from the data available. In most cases this leads to a significant improvement in the *status quo* (for example, higher profits). In addition, companies are – quite rightly – extremely cautious in using personal data.

The good news is that to establish best prices at product level it is not necessary to elaborate personal data.

Lastly, the data gathered must be cleansed of errors and prepared for further elaboration.

This is a demanding step, because data in various formats from different sources must be brought together. The task should therefore be carried out by experts, data scientists, to ensure that the data is correctly and completely transformed into an algorithm.

2. Training of the algorithm

The next step is to "train" the machine-learning model. At first, the model analyzes all the variables and determines the possible effects of price variations on sales. In doing this, the machine-learning model automatically detects the correlations and models that human analysts may easily overlook. These are incorporated into the algorithm to calculate best prices and constitute the basis for sales and profit forecasts. Once created, the initial model is submitted to a practical trial and can also be manually optimized on a regular basis. After each correction, the algorithm learns and improves its results independently.

More sets of data can be added to further improve the accuracy of the algorithm. Over time, the training effort diminishes, whilst the efficiency of the software continues to increase.

3. Optimization based on forecasts

Once developed, a machine-learning model can set best prices to satisfy specific company objectives and determine elasticity of pricing for thousands of products in just a few minutes.

The internal marketing and sales teams can use these calculations to experiment more boldly with entry prices and markdowns, because they can better judge the potential impact on sales and on demand.

Instead of relying on instinct and experience, they can now reason on the basis of the results of the machine-learning algorithm. This gives them room to maneuver, which generally results in an increase in sales and profits.

Applications of AI-Based Pricing: B2C and B2B

In the retail sector, AI-based pricing will become increasingly mainstream over the next few years. According to a worldwide study by the management consulting company Horváth, 79% of retail companies foresee investing in AI by 2030, partly for the purpose of optimizing prices.

Many well-known retailers are already exploiting the power of machine learning. They include some famous brands such as the US electronics company Monoprice, the British supermarket chain Morrisons, or Zara in the fashion sector. In the latter case, for example, the Spanish fashion chain determines its entry prices by AI and lets prices react to trends automatically. As a result, Zara only has to sell 15–20% of their products at a discount according to Ghemawat and Nueno, as against the 30–40% of other European retailers.

Ralph Lauren and Michael Kors instead are using machine learning to sell fewer items on discount, manage their inventories and increase sales.

Boohoo and Shein – fast fashion retailers – are well known for using machine learning in order to reach their business objectives, despite their low entry prices.

Even in the B2B field, an increasing number of companies are introducing AI-based pricing management.

Let us take a look now at six applications of AI-based pricing according to type of company, both in the B2B and in the B2C context (see Figure 7.2[5]).

Geo pricing

In several geographical areas within a nation (Germany: north, center, south) or a geographic region (Europe) willingness to pay may vary, even considerably. AI makes it possible to combine internal sales figures, product, and customer with external data – for example, demographic, income and economic – indicating by means of geographical pricing, "geo pricing" for short, the ideal price-point per product, service, or discount.

Contract margin predictor

In selling contracts, whether for maintenance or post-sales assistance, the price or discounting may vary a good deal according to the type of customer or the seller's negotiating power.

So as to allow only the "necessary" discount for closing the contract, AI analyzes a series of parameters linked to the contract, and others linked to the customer and context of transaction, in order to indicate the price and profit margin that are best suited for closing the deal.

Churn minimizer

In many industries, characterized by an ample customer base, such as phone companies, pay TV, electricity providers, or post-sales

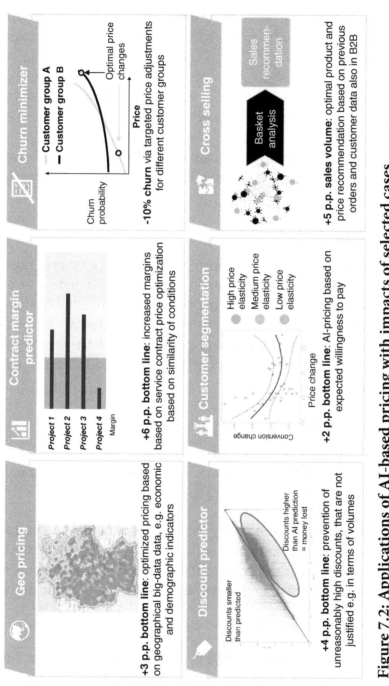

Figure 7.2: Applications of AI-based pricing with impacts of selected cases

Source: Courtesy of Horváth

assistance for automobiles or machinery, minimizing customer churn rate is essential for the company's success.

AI makes it possible to create indicators that will calculate churn probability and suggesting retention measures for preventing clients abandoning the customer base, also coming up with the optimal price to propose to customers at risk.

Discount predictor

A typical case of application is that of optimizing discount. Starting out from the price list, the algorithm indicates the maximum discount necessary for closing a sale, without going beyond the necessary concession. The optimal discount is predicted for a given customer and a given deal.

Customer segmentation

To segment customers several paths can be pursued. A typical approach is to evaluate willingness to pay. AI, for example, is well suited to calculate present customers' willingness to pay in order to segment them and optimize the offers to be made to them.

Cross- and up-selling

Both in the fields of B2C and in B2B an attempt is made to sell a customer an accessory or additional service to increase revenue and, at the same time, profits.

AI makes it possible to analyze efficiently which combinations of products have been sold to clients with similar characteristics, to make a direct offer – in the case of a B2C customer – or, for example, through the retailer in the case of B2B – of further products and encourage cross- and up-selling, also coming up with optimized prices.

Summary

AI-based pricing regards the use of methods such as artificial intelligence, machine learning, and deep learning to imitate human behavior and take autonomous pricing decisions that improve constantly, thanks to advanced statistical methods and algorithms.

The steps for setting up an algorithm for optimizing prices based on machine learning are fairly simply:

1. Collection and cleansing of data;

2. Training of the algorithm;

3. Optimization based on prediction.

Some applications of AI are:

- Geo pricing;

- Contract margin predictor;

- Churn minimizer;

- Discount predictor;

- Customer segmentation;

- Cross selling.

AI-based pricing can have a strong positive impact on business success and this is why it is spreading increasingly, both in the field of B2C and in B2B – with B2C being more mature than B2B.

To stay competitive, it is therefore necessary to evaluate how to keep up with the competition on the front of AI-based pricing, too.

CHAPTER 8
FREEMIUM

"The purpose of business is to create and keep a customer".

Peter Drucker

Case History

For years you've been waiting for that new album by your favorite singer, dreaming of the vintage LP humming on the turntable, 10–12 new tracks and you, sitting in your favorite armchair, the house quiet, maybe with a glass of good red wine to hand. You'd pay anything to be amongst the first to have it once it's out. And then, finally, unexpectedly, the moment suddenly comes: the album is released! You're excited. Curious. You get up and shower, get dressed. You watch a scene from *Saturday Night Fever* once again on YouTube. You do a couple of dance steps like Uma Thurman in *Kill Bill*. You're in gear, ready to go BUT (there's always a but) the hitch comes: you don't have to buy it, or go out, or find somewhere to park or even go inside the record shop you've already seen on the corner, which reminds you so vividly of *Championship Vinyl*. And you won't even have to pay the price. . .why? Because someone gives it to you as a gift![1] And there it is, the new album by your idol is right there in front of you with the plastic cover still pristine, and it's free! A dream has come true and it isn't even 8 o'clock in the morning yet.

When Prince's new album, *Planet Earth*, was released in the UK in 2007, almost 3 million people got a copy.

Normally news like this evokes images of big shots in the record industry giving one another the high five and fans pouring into the shops to empty the shelves. But in this case none of this happened: Prince didn't sell one (!) single copy of his album, evaluated at around $20, in the UK.

With an unprecedented gesture, Prince decided to distribute the new album free with the British tabloid *Mail on Sunday*.

Readers who paid the usual price of around $3 for the newspaper received the album free.

Weighing up the monetization of the album alone, Prince certainly didn't make any money: instead of receiving the usual sales commission on sales at full price – in this case around $2 – through the traditional channel, *Mail on Sunday* paid only 36 cents per copy as license fees.

But considering the impact this operation had on ticket sales at his 21 London concerts, for *The Purple One* it was certainly well worth it! In Great Britain alone, sales of all the available tickets were an absolute record for this sadly missed artist.

Although at first sight, Prince was giving up a commission of $4.6 million, the promotional effect on the 21 concerts generated revenue worth $23.4 million, with a profit of $18.8 million.[2] Added to which, sales of the *Mail on Sunday* increased by 600,000 copies more than the average 2.3 million, that is, a rise of more than a quarter of average sales in just one day.

Although this did not cover costs, the newspaper itself considered the promotion a great success; the reason was partly because by positioning themselves as a market innovator through such a praiseworthy and groundbreaking operation, *Mail on Sunday* became more attractive to advertising companies.

In this case a *loss leader strategy* is spoken of, that is, money is lost on one product (in this case the album), so as to sell other profitable ones (the concert tickets).

To sum up the concept, you only have to think of a pub: salted peanuts are offered free, so as to make customers thirsty and sell expensive drinks, such as beer, cocktails, and all that makes up the pub's core business. The small generates the large and David defeats Goliath.

This is the same principle followed by Google, which for years has been successfully selling at (more or less) zero; hundreds of products are provided free of cost: as well as offering the world's leading search engine worth a market share of well over 90%, Google offers services in various contexts from email (Gmail) to information (Google News), navigators (Google Maps) to translators (Google Translate), to the sharing of documents, spreadsheets and images (Google Docs).

Advertising revenue and a few other sources of income are so high that Google is able to offer a considerable number of free services.

When Google's product managers wish to launch a new product or service, they don't ask themselves how much revenue this will generate so much as whether the offer will be appreciated and if/to what extent it will be adopted by a large number of users. This represents the basis of Google's strategy: to launch services of interest to the widest possible customer base and spark off mass adoption, so as to sell advertising.

By selling completely free products Google became a 182-billion-dollar giant in 2020, with profits reaching $40 billion, a sum that comes to more than the profits of all the American automobile and airline producers put together.[3]

This was also the direction taken by Michael O'Leary, who as CEO completely transformed Ryanair, making it into one of the most successful low-cost airlines in the world. O'Leary maintained that in the future he would be able to offer his customers more or less free flights. The revenue would come from sharing income with the airports.[4] Naturally, there may be problems if you saturate the market or if your competitors start regarding you negatively. The same can happen if the skies are saturated – up until the coronavirus swept them clean, leaving human beings with clear horizons and white clouds floating in a blue sky. The 2020–22 pandemic certainly slowed down the vision, but what is certain is that "everything flows" in life and in the same way the *freemium* model will continue on its path in many sectors, overcoming this umpteenth obstacle.

Analysis of Context

"Freemium" is a combination of the word "free" and "premium" (or higher amount).

From this first semantic analysis then, we can see that the word describes a pricing strategy in which the initial function is provided free of charge with the possibility of accessing other "connected" functions on payment.

Like a sort of game in which, in order to unlock the next stage, I must first complete the previous one by exchanging the doubloons I've managed to get hold of up to then. Fans of the 1980s/90s will remember game upon game of *Golden Axe* at the arcades with the big screens and game sticks, playing against trolls, werewolves and gnomes and scaling the highest of walls, accumulating lives and refining their spirit of adventure whilst, why not, making a good bit of money before going on to the next screen.

The objective of the freemium is to initially draw – or better, attract – the largest possible number of customers by means of a free offer and then, once users are familiar with the base function, the supplier hopes that their willingness to pay for added, more sophisticated services will increase.

In this sense, the freemium can also be interpreted as a form of penetration strategy, by which a sort of cross-subsidization occurs between free products and those sold at a price that will cover the cost of the free offer and generate a profit margin. In particular, three forms of cross-subsidization can be distinguished.

1. Free

The first case regards free products directly subsidized by products paid for. A good example, in this sense, are the classical "pay for two and get one free." This is how Trony in the field of domestic electrical appliances offers one product – the least expensive – free when purchasing three items.[5] Walmart makes the same offer for DVDs: one is free and the other is paid for, with the objective of drawing customers to the retail point and having them fill the shopping basket with profitable products. Or Vodafone and O2, who in Germany give you a smartphone free if you sign a two-year contract with them, which also covers the cost of the mobile phone. And again, the bank Unicredit offers a free credit card, so as to monetize the current account sold together with the free card.

Compared to the case of the *loss leader* – such as the Play Station games accompanying Sony's console and sold below cost, or the expensive wine in the restaurant that subsidizes a low-cost menu – at least one of the products comes completely free of charge.

2. Freemium

The second case regards products that are free as a base version and subsidized by advanced versions which are paid for, and this is the case of the real freemium.

Whilst in the first case, the acquisition of the free product is in fact bound to the purchase of a product that is paid for, in this case the free product can be used without any further buying.

How?

Freemiums are particularly suited to so-called "experiential commodities," that is, those whose full benefits only become manifest when experiencing the commodity.

Consumers with previous direct experience of the product tend to show stronger connection and intention to purchase than consumers who arrive through indirect channels.[6]

This is true of software products such as Adobe, which grants a base version of its software free of charge and a professional one against payment. Or, for example, social networks like LinkedIn or Xing, where many functions are free (for example, creation of a profile or sending messages) whilst others have to be paid for.

The digital era has allowed these models to spread more rapidly, so that they are now widespread on the market for digital products.

In the offline world, where freemiums were represented until a short time ago by samples of perfumes in perfumeries or sample tins of food given to customers to encourage sales, costs were such as to limit their use; on the digital market, instead, costs are marginal and insignificant, so that opportunities flourish. If only 5% of users were to pay for a product, the business model would still hold out. This means that, even if 95 users out of 100 didn't pay, the remaining 5 users would be sufficient for the company to generate profits, thanks precisely to the negligible costs involved.

3. Triangulation

The third case regards the bipartition of free products with products which instead are paid for.

This is the classical model characteristically used by the media: in the free exchange between two parties a third party is introduced, which is paid for. This happens with *Metro*, the United Kingdom's highest-circulation print newspaper: the newspaper is distributed free of charge and offers readers news and articles of clear interest, so that they have an inkling of the main events of the day. It is a horizontal read, without detailed articles, and in fact we are often short of time, so *Metro* exploits our skimming abilities.

In any event, the important thing here is the price the product is sold at: the third party in this case is the advertiser who pays the publisher the fee for ads published in the newspaper. Thus the publisher doesn't so much sell copies of the newspaper to readers, as readers to advertising agencies; this is where a sort of *triangulation* steps in.

The same is true for TV or radio stations. On the Internet, entire media ecosystems based on this *triangulation* mode have grown up, which are based on: (a) free content; (b) sale of information related to users; (c) pay subscriptions.

In reality, further examples are also to be found offline, for instance in the case of credit cards: American Express gives its card free of charge to users and takes a percentage from the trader.

And now for a mathematical consideration that seems more difficult than it really is. Let's take a look at it together.

The three types of cross-subsidization are actually based on two types of pricing:

- In one case the price is equal to zero, that is, there is no payment and the transaction takes place free of charge.

- In the second case there is a price and it is linked to the compensation.

It is possible to find a third case: negative pricing. In the latter case the consumer is paid to use the product and not vice versa. One example is Microsoft who pays users to carry out searches on Bing, allowing them so-called *rewards* that can be converted into various premiums.[7] Or airlines' customer loyalty programs, such as the AAdvantage or Miles & More: the latter, for example, makes it possible to acquire goods and services by paying for them through Miles-Pay with miles instead of dollars.[8] Another example comes from the cash-back formulas, from automobile companies and dealers like General Motors or Chrysler. The offer from a chain of Danish gyms is oriented in the same direction: if you go to the gym at least once a week, your monthly subscription is free. Every week you don't go (and we all know what this means after the first month when motivation is high, how much it "costs" in terms of willpower to persist, with that annoying, insignificant yet fatal ankle pain), we find ourselves charged the normal monthly membership rate.[9] A great incentive – you might say they do it for our own sakes.

In fact, when you find yourself paying because you weren't "active" enough at the gym, the designers of this formula will certainly be thinking, you'll berate yourself for failing to go; not actually your fault of course, it's always someone else's fault when you don't go and you'll be swearing to be more consistent in the future. If, instead, you pay the yearly subscription but from time to time you fail to go, you'll tend to ask yourself whether it might not be as well to cancel the subscription: instead, here is a pricing model that is more likely to keep this type of customer.

In Los Angeles, if you're in a band, sometimes it isn't the venue that pays the band, as much as the band that pays to play in the venue.[10]

This is often the case for new emerging groups aiming more for fame and visibility than for cash. Once they have a name, then they'll be able to turn the tables and finally get paid.

In the sector of telephone conferences there are several players who offer their services to users free of charge. These are companies such as FreeConferenceCall, used by more than 42 million people, including mainly companies[11]: the service is free to the user, as the telephone companies pay a commission on the international calls made by users to take part in the free telephone conference.

All these cases show how smart companies have succeeded in inverting normal cashflow by offering a free product.

Let's get back to the freemium, which deserves closer attention.

This particular business model has taken root in software industries, where software companies like Adobe advertise their software in free "lightweight" versions. The growing popularity of the freemium has caused this approach to monetization to spread in several contexts.

A successful practitioner of freemium prices is SurveyMonkey, which acts as follows: if by any chance you might wish to carry out a quick online survey with a maximum of 100 interviewees, you can draw up a questionnaire and use their survey service free of charge. Using this strategy, SurveyMonkey has attracted over 20 million users.

Naturally, the profitable business depends on something entirely different: a group of customers who want replies from more than 100 people and are therefore willing to pay a subscription for SurveyMonkey's advanced services.

Other cases of freemium pricing are to be found in widely varying environments:

- *Flickr*, for example, offers free space for sharing photos and videos but sells extra space for archiving;
- *Skype* offers free calls between computers but sells calls between computer and phone;
- *Fortnite* offers a free game (*Battle Royale*) up to a certain level but sells accessories (for example, skins) and subsequent games (like *Save the World*).

Many other examples are to be found in video-streaming (Youtube, Vimeo), music-streaming (Spotify, Deezer, Pandora), cloud storage solutions (iCloud, DropBox, Google Drive, OneDrive) or social media (such as Xing and LinkedIn) – all these projects have this type of pricing in common. The freemium strategy is not, however, limited to immaterial goods on the Internet.

Vistaprint, a Dutch multinational of French origin which uses mass personalization to produce short-term print jobs – such as cards or flyers – has for years used its "free visiting card" promotion to encourage customers to order from them. Whilst the company gives away millions of free print jobs (of course covering the costs, at least partly, by the surcharge on "delivery and handling") it generates over a billion dollars a year in paid print jobs.

Apparently freemium models existed in the old world, too. For example, for years the banks advertised current accounts without commission. Only if the customer demanded extras on top of the base service, did they have to pay. Nonetheless, the free base account was generally linked to certain conditions: for example, the account had to have a minimum balance. In the end the customer paid with the interest lost. The same is true of so-called "zero per cent" financing, increasingly offered by retailers over the past few years: in reality, here, the costs of financing are hidden in the purchasing price. Another difference comes according to whether the free offer is linked to advertising or not.

In many freemium services the offer really is "free" in the sense that no advertising is shown. One example is the Microsoft Office version for smartphone or tablet, a basic version of which is made available to customers free of charge. At the most, the user "pays" with their data. LinkedIn and Xing are further examples.

For other services, the user must accept advertising or commercial breaks in his "free" offer. This is the case on Spotify, where users of the premium version pay $9.99 a month and receive the music without commercial breaks. The free customers have to put up with the advertising, as they would on the radio. They "pay" with their attention. The same goes for YouTube.

Another example is to be seen in the Italian newspaper *repubblica.it*. Although several articles are freely available to online users, certain selected articles can only be viewed after having subscribed to the website. Two options are available to users. One is the 6-month digital subscription to the site only, which is available for €1/3 months and then €5.99/3 months, and the other is a 12-month, online subscription for the website, daily and supplements, marked by a price of €5/3 months and then €13.99/12 months.

The dictionary providers leo.org offer their services free of charge but show advertising. If you block the latter a request for a donation appears which can be interpreted as a variation on the pay-what-you-want model, dealt with in another part of this book.

LinkedIn goes further, differentiating prices according to different demands. The premium career offer, aiming to "land the job of your dreams," costs $29.99 a month. For the premium business offer, which serves to "select and nurture your network," the monthly fee is $59.99. To "unlock sales opportunities" the price rises to $79.99. Obviously, LinkedIn estimates to different extents the willingness to pay of candidates, those interested in networking and sellers. In addition, there is a good 20% discount on all three offers in the case of annual subscriptions.

Xing, too, has a freemium model. The communications software Skype is based on full functionality, but limits free calls to its own network. Once users get used to intuitive interfaces, they are more willing to pay a price to call landlines or mobile phones. At the start, Skype mainly sold minutes of individual calls. Later, the offers were structured similarly to those of the traditional telephone companies. The present pay offers include packages of minutes or flat rates towards selected national networks.

4. Rules for successful freemiums

All "free to play" videogames are good examples of the freemium model. They can be considered an "Extreme wall" – that is, we no longer expect 80% of profit to come from only 20% of customers – rather, the business model depends on attracting millions of players, only a fraction of whom will make purchases "while playing." For example, at the time of Zynga, the creator of the popular online game *Farmville*, a study by the *Wall Street Journal* found that fewer than 5% of the players had bought something. . . not even a bale of hay costing $1 for the virtual cows (!).

This application of the freemium pricing strategy thus becomes vulnerable to the whims of a fairly limited number of players who may be attracted by something else, for example the next new game. Even more so than in the case of flat rates, it is important to the freemium models for marginal costs to be equal or as near as possible to zero, at least for the base service, and thus for the "free cost" not to create a burden for the supplier.

There follow the four rules for making the freemium pricing model efficient.

1. The market must be segmentable

For the freemium model to be successful, we need various market segments with groups of customers who are looking for different benefits. If more or less all the customers look for the same characteristics in a product and the same level of performance, then the freemium model will not work. This is why Facebook works, thanks to support from advertising, whilst LinkedIn has had success with the freemium model. Almost all those on Facebook are looking for the same characteristics and benefits. Instead, on LinkedIn most of the occasional users make use of the online service free of charge as a way of keeping up with their work contacts and to host online a brief summary of their professional bio. But the head hunters want more. They will willingly pay a monthly subscription to be premium users of the service. The distinction is that the premium users can contact any candidates they identify as possible correspondents for an opening, whilst ordinary users can only approach other people on the service through presentation by someone with shared links.

2. The product must have a low variable cost

For almost all websites the marginal cost of an additional customer (think of the 800 millionth customer on LinkedIn – who joined in 2022) is close to zero. The complete cost of 100 visiting cards on

Vistaprint would be fairly high if we were to assign all the general costs to it, including the staff at the central office and factory operations. But, given that the production and general trading functions of the company are covered by pay orders, the cost of one order beyond the margin is quite small. Instead, neither a mechanic nor an optician can successfully use the freemium model because in any case the costs of serving one extra customer are a long way beyond zero.

3. Freemium customers should act as ambassadors for the pay version

Dropbox, a cloud-based file-hosting service, deliberately recruited users to upload personal files with the certainty that they would demand a file-storage system that was also cloud-based and that could thus be sold to companies at a sufficiently high price to cover the entire operation. Between consumer customers and businesses the market was easy to segment on the basis of the total storage quantity that each would use.

4. Gradual addition of restrictions to the free version should be introduced

When SurveyMonkey started out, the only restriction was the number of surveys that could be collected. The more success the service gained, the fewer were the functions of the pay service available to freemium users, since the website limited the cutting and pasting and the saving of results as a .pdf (even though the results could be visualized online). This was highly successful in convincing an increasing number of users to opt for one of the company's pay offers.

To trigger paid usage, different types of possible limitations to the free version can be established: (a) *functionality*, as in the case of LinkedIn, where the base version makes it possible to do certain things free of charge while the professional one instead demands some kind of payment, in return for more utilities; (b) *temporal*, as in the case of Salesforce, which allows 30 days' free use before asking for payment; (c) *use*, as in the case of Intuit QuickBooks;[12] and (d) *type of customer*, as in the case of Microsoft for StartUps which offers Azure free of charge[13] to companies less than five years old and with a turnover of below $10 million.

Summary

Freemium is a pricing strategy. The initial functions are provided free of charge and later they are "unlocked" on payment. The objective of a freemium is initially to attract the highest possible number of potential customers by means of the free offer.

Once users have become familiar with the base functions, the provider of the service hopes that their willingness to pay for added services of higher quality will grow step by step with the offer.

In terms of cross-subsidization to support the free-offer model, three distinctions are made:

1. *Free*: the first case regards free products directly subsidized by pay products. A classic example is the "buy two and get one free" offer.

2. *Freemium*: the second case regards products whose base version is free and subsidized by advanced versions of pay products, that is, the freemium.

   N.B. Whilst in the first case the acquisition of the free product is linked to a pay product, in this case the free product can be used without having to pay for something else.

3. *Triangulation*: the third case regards bipartition of free products, in return for products paid for. This is the classic model that characterizes the media: in the free exchange relation between two parties, a third is introduced which is paid for.

The three types of cross-subsidization are based on two types of pricing: in one case the price is equal to zero, as there is no charge, in the other case price is based on payment. In reality it is possible to come across a third case: negative pricing. In the latter case the consumer is paid to use a product and not vice versa. One example is Microsoft paying users to carry out searches on Bing, allowing them so-called *rewards* that can be converted into different premiums.

The four rules for making the freemium pricing model work efficient are:

1. The market must be segmentable;

2. Products must have a low variable cost;

3. Freemium customers should act as user-ambassadors of the pay versions;

4. Gradual addition of restrictions to the free version should be introduced.

It goes without saying that the freemium strategy can only be successful when there are enough users of a product's pay version to allow the business to break even.

In the future many companies will have to adapt to a direct competitor offering free goods and services.

CHAPTER 9
SYMPATHETIC PRICING

"Life is what happens to you while you're busy making other plans."

John Lennon

Case History

You're on your way back from work after a tiring day. You can't wait to get home.

But when you get to the station, you realize something's wrong; people walking up and down, distracted and disoriented, talking nervously on the phone, children crying, girls sitting on their backpacks smoking, their make-up starting to run.

The platform is full of life, though chaotic: men with moustaches and briefcases distracted by repressed desperation, an unhappy fate.

Public transport has been hit by a general strike. There are no more trains running.

A window of nightmare opens up on daily life.

A whole lot different from the *best of all possible worlds* you'd been imagining: a hot shower and music, comfortable clothes and a glass of good wine, sitting down and read the morning paper you hadn't even managed to open yet.

This is how the idea of domestic bliss is, if not completely shattered, at least delayed.

Reality surpasses fantasy.

So here we are, sweating, shirts sticking to our skin, armed with a patience that threatens to elude us.

A Day of Ordinary Folly, like Joel Schumacher's film with Michael Douglas. And instead, take a deep breath, again and once again. There's nothing else to be done.

And instead. Something unexpected happens.

Just as you're imagining the nightmare, convincing yourself that you're doomed to spend the night in a crowded waiting room, comes a pop-up from Uber on your mobile phone, telling you that, as a contribution to relieving the crisis, you are entitled to a 50% discount on a "ride home".

"Life is what happens to you while you're busy making other plans," and John Lennon was right.

Perhaps that's what life is all about: the unexpected turn of events; something that cannot be put into order.

You think you can control everything. And instead, *life is what happens to you while you're busy making other plans*.

So this is the key to everything, then.

The message that can be transferred to the customer.

In fact, after a comfortable and inexpensive car ride (the crowded train just a bad memory), there you are at home, safe and sound.

Even better than usual, the customer will think, forever grateful to this unexpected and fortuitous "gift" of destiny.

The example may seem unreal and implausible, but instead it is exactly what Uber offered their customers when public transport strikes hit both Boston and London.

This is sympathetic pricing, whose objective is to transform the customer's negative experience – in this case the transport strike – into a positive one – that is, the inexpensive ride home – creating empathy with the brand, the idea, the suggestion at an unconscious level, generating a positive attitude and emotion in the customer towards your brand.

> *Sympathetic pricing* can thus be defined as:
>
> *The application of flexible and imaginative discounts that help relieve peaks of pain in lifestyle and lend a hand in difficult moments or uphold a shared value.*[1]

Analysis of Context

Sympathetic prices can make the difference to customers in a whole series of situations. Applying these "social" prices does not have an immediate positive impact in the timeline of the company's objectives but finds its place in a correct logic of return in the medium and long term.

If the 1950s and 1960s saw the stuff of dreams make its début on the world stage, and the 1980s the peak of easy money, after the 1990s – and even more so the year 2000 – came the burden of uncertainty: in

September 2021. Twenty years have passed since 9/11 – the terrorist attack which wiped out the Twin Towers in New York, leading us into an era of surveillance capitalism as necessary as it is dangerous for all that regards individual freedom and contemporary democratic societies.

Although companies continue to tell their customers that they concentrate on their needs and care about them and their daily challenges, they are not believed. A recent research study by the PR agency Cohn & Wolfe shows that only a small minority of consumers trust companies: only 5% of consumers in the United Kingdom and the United States believe that big companies are really transparent and honest.

Dozens of reports, opinion polls and consumer studies confirm this data: when it's a matter of really concerning themselves with their customers, of possessing a higher objective and, in general, being a more human brand, people generally think that most companies haven't yet reached this point.

There still exists deep skepticism in many consumers even after years of campaigns, messages, and work aiming to demonstrate that the brands really do care. Usually, these initiatives are seen as business oriented or, at best, relegated to the area of vision and vague promises, until they vanish in a sort of *white noise*: consumers have learned to take them as a fact or simply to ignore them.

According to another recent study, fewer than 10% of US consumers and 20% at a global level think that the brands "really do make a difference to people's lives."

Customers would seem far more inclined to embrace the few brands that are already trying out a new, bold approach in attempting to become more human, or those that are committed to a new approach to flexible pricing.

Both in the B2C and in the B2B sectors, sympathetic pricing helps to improve the image and perception of companies that practice it by increasing confidence in them. Offering discounts at the right moment is the most effective proof that a company cares about its clients. This will translate into a better image for the company, the recovery of customers' trust and new, precious brand supporters.

In the long term, this, together with possible positive media coverage, will translate into a baseline increase.

Companies should therefore look carefully at how they use their sympathetic pricing.

Let us move on to an analysis of the three types of possible applications of sympathetic pricing[2]: (1) painkiller pricing; (2) compassionate pricing; and (3) purposeful pricing.

1. *Painkiller pricing*: companies use this in its literal sense of "painkiller" to help their customers overcome everyday irritations. The Uber short story is a typical example of painkiller pricing, but there are many others to quote.

   For example, "Did you get a parking fine today? Then, you deserve a free drink!": this is what the Australian restaurant *Melbourne The Wolf & I* offers every Thursday at their Local's Night, to anyone who has been given a parking fine.

   And some time ago, the Argentinian domestic electronics brand BGH launched a special summer campaign called "My house is an oven": discounts on air conditioning were given to people whose homes were considered too hot. Customers had access to a website which helped them trace their apartment's exposure to the sun. The greater quantity of solar rays their home absorbed, the bigger discount they got on a BGH conditioner. The campaign was run in Argentina over the summer months – which at that latitude last from December to March – and in the period considered it brought home over 49,000 sales. According to the famous advertising agency Saatchi & Saatchi, this form of pricing has earned over $14 million in sales for the company's balance sheets.

The slogan "Bald is beautiful" is displayed in pride of place above the Japanese restaurant *Otasuke*, in Tokyo's Akasaka neighbourhood in the city center, where customers with "follicle complaints" are welcomed with open arms and offered discounts which are not allowed for their "hairier brothers and sisters."

Although not widespread in Japan, baldness affects 26% of men, according to Aderans, an important Japanese wig producer. Genetics plays an important role but also the stress of company "salarymen" with chronic workloads. This is why *Otasuke*, which in Japanese means "helping hands," has translated its prices into a *sympathetic* formula, so as to generate discounts for bald customers. In this way they encourage their customers to accept their loss (of hair) and not hide it. This also happens in the precious art of *kintsugi*, or the century-old practice of repairing broken chinaware using gold. *Kintsugi* focuses on the cracks and values them: literally, the term derives from "gold" ("kin") and "unite/repair/reunite" ("tsugi").

The metaphor is clear: the idea is not to hide the wound (in the object) but to remake it, giving it a new form that takes its inspiration from the old one but with an extra vein, enhancing in some way the change it has undergone. In the same way, "Baldness is a delicate matter in Japan, yet in Hollywood there are stars who ignore the problem and proudly pursue their careers," says the owner, Yoshiko Toyoda, "I thought it would be a good idea to promote this sort of spirit.[3] A sign outside the venue, explains his support for "our hardworking fathers, who lose their hair," because of stress at work. Every bald customer receives a discount of ¥500, around $5, and the premiums increase with the number of bald customers in each group. If five go and have a drink together, one drinks free of charge. The posters on the pub's walls narrate interesting details about baldness (*Which nation has the highest rate of baldness?* The Czech Republic, with 43%, followed by Spain and Germany).

Another interesting case is that of discounts on hotel rooms when it rains: the Noosa International Resort came out with an offer called *Rainy Weather Rebate*, consisting of a 20% markdown on the price of a room, if local rainfall was over 5 mm during a visitor's stay. Situated in Queensland, Australia on the famous Sunshine Coast, the resort came up with a scheme for attracting tourists after a sudden spell of unseasonal bad weather (which in this case included four cyclones, heavy rain, and floods). Taking into consideration this "collateral damage" in the midst of *Antropocene*, an era increasingly marked by considerable instability, in terms of the climate too, can be highly strategical. It gives our customers an idea of how well informed and sensitive we are to the big challenges of modern life.

2. *Compassionate pricing*: companies offer support to their customers with a message that might be summed up as, "*When life is not treating you well. . . we'll take care of it!*" And companies offer discounts or free services.

Sufficient to think of what happens with regard to discounts on food products for those who are living below the threshold of poverty: Community Shop, a British supermarket, applies compassionate pricing by selling discounted brand products to people on welfare. The project is supported by high profile chains and brands such as Asda, Marks & Spencer, Tesco and Tetley, which supply the shop with products that would in any case not come up to their standards. These stocks would generally end up in tips or be transformed into animal food.

Another compassionate pricing initiative was created by *Tienda Amiga* in Spain's capital city, Madrid: in this case, small businesses offer discounts to the unemployed in their neighborhood. The name of the company means "Friendly Shop" and it was created by the *Asamblea Popular de Hortaleza*, a community group in Madrid. Their objective was to establish a more ethical local economy. In just a few months over 150 local shops joined, offering discounts of between 5% and 20% to people without jobs.

Access to free databases for journalists who had lost their jobs: this was how Pressfolios offered the service free of charge to journalists who had been laid off by *Star-Ledger*, the biggest newspaper in New Jersey. Pressfolios allows users to create an online portfolio. The journalists made redundant receive a free Pro account, with unlimited storage for three months, which would normally cost around $15 a month. Small steps for a company, but a great help to those who have lost their position and certainties.

Lowe's manages or provides support for over 2,370 shops selling household articles and ironmongery in the United States, Canada, and Mexico, with over $65 billion of sales before the pandemic. The Canadian branch of the chain introduced compassionate pricing when a winter storm destroyed most of the trees in Toronto. Lowe's Canada gave the city 1,000 red maples. The trees, worth CA$30 each, were available free of charge in the parking lots of two of the shop's Toronto locations, and were distributed on a first-come-first-served basis until supplies ran out.

Groupon India, which manages an online market for local trade, putting traders in touch with consumers and offering goods and services at discount prices, applied compassionate-price services in response to the increase in the cost of onions. Through the daily offers on their website, buyers could buy 1 kilo of onions for 9 Indian rupees (US$0.15), corresponding to around one-eighth of the main price, including home delivery.

3. *Purposeful pricing*: by means of purposeful pricing companies help groups of people with shared values and lifestyles, offering discounts, freebies or reductions.

It is purposeful pricing that made it possible to recruit the elusive Millennials – young people between the ages of 18 and 35 – who seemed to be more concerned that chocolate bars were cheap than about their history.

Recruiting this age group is essential to brands for protecting the company's future value. In this context, another market challenge is the growing predominance of retailers who demand stricter trading conditions. The two main Australian distributors, Coles and Woolworths, for example, demand exclusive product lines, significant business investments, and reduced prices to even admit a brand into their product selection.

This behavior is becoming more and more common in the convenience channel.

Here, the case of the Mars brand Snickers is emblematic, as they needed their history to carry greater weight. It was also essential to achieve this in a way that would unlock new sales opportunities in the retail sector, to contrast the pressure on prices and convenience.

The challenge was to develop an idea that would unleash sales demand for a retail sales partner whilst at the same time guaranteeing that they would tell the story of the brand. All this with a public that was difficult to contact, rather indifferent to marketing activities, always on the move, and online.

This all became possible thanks to the creation of *hungerithm*: a hunger algorithm that monitors customers' mood.

The algorithm was constructed by taking into consideration a lexis of 3,000 words, analyzing over 14,000 tweets a day.

Every tweet was reinterpreted as a single data-point capturing polarity, subjectivity and intensity of language. Transforming the tweets into data-points it was realized that there was no need to build a database or ask questions of consumers.

Every 10 minutes a set of combined data was analyzed and consequently assigned to one of ten different, predefined states of mind. Every time the sentiments of a cluster of users studied on Internet was "cool", Snickers stayed at A$1.75. But when the result was "off their heads," prices might drop to A$0.50!

Hungerithm operated live 24 hours a day, 7 days a week for 5 weeks. People could monitor the trend online (www.snickers.com.au), request a Snickers coupon directly on their phone and cash it at a nearby *7-Eleven* convenience store. This targeted pricing approach was an involving and up-to-date way for Mars to talk to Millennials and increased the sale of Snickers by almost 20%. A quite satisfying result in the stagnating category of chocolate bars.

A second good example of targeted pricing was the campaign by the Dutch airline Corendon, upholding the gay rights movement in Russia during the XXII Winter Olympics in Sochi. The airline offered passengers who supported gay rights a 50%

discount on the price of air tickets. These were available for a one-month period at prices ranging from $399 to $799.

RATP, the French public transport company, offered free or discounted public transport on days at risk of smog, in order to fight pollution, thus encouraging citizens to leave their cars at home. The first offer was made by RATP in response to dangerously high levels of pollution that enveloped the French capital for days, something that is unfortunately becoming common in many of the world's cities.

Easy Taxi, one of the most frequently downloaded taxi apps in the world, founded in Brazil and now available in 12 countries and 170 cities, launched a pricing initiative that foresaw waiving the R$70 booking fee for every lone female passenger. The move was planned to encourage women to continue using taxis following a series of physical attacks by taxi drivers on their female passengers.

AnchorFree, a software company in Silicon Valley, made *Hotspot Shield* – its app protecting online privacy and costing $140 for an unlimited subscription – a free download in Venezuela. The application of the software brand based in the United States allows users in the South American country to get round the government blocks on the Internet. The offer came in response to the Venezuelan government's growing censorship of the Internet, stimulated by increasing civil disorder.

In a far less political but more social fashion, cafés and restaurants have started to offer discounts for smiling or to encourage customers to say: "Good morning," "Good evening," or "Please"; this happens at the Petite Syrah café in Nice on the French Riviera, which offers a discount to customers if they pronounce the magic words "hello" and "please."[4] A cup of coffee with a "Bonjour" and a "S'il vous plaît" costs €1.40 (around $1.50), whilst customers have to part with almost $4 more if they fail to make the greeting, as much as $8 more if they also omit a minimum of socially acknowledged courtesy and address the barman curtly, demanding "a coffee."

Up until now we have concentrated on B2C cases. In the same way, we should consider that sympathetic pricing strategies – *painkiller pricing, compassionate* and *purposeful pricing* – are also widely applicable in the B2B industry.

The reason is that company buyers and members of govern-ment sourcing staff depend on a professional mind structure, which in some way will affect (positively as well as negatively) their personal life. We are tightrope walkers. Our lives are bal-anced between inside and what is to be seen outside. The center is what we manage to carry between one bank and the other of a river that is no more than our own point of view on things.

People have the same skeptical point of view before the pres-sures of company marketing that advertises green products (often only superficially green, which explains the term "green washing") or socially responsible ones, substantially the meas-ure of how involved the company in question is. The price sells, giving rise to the idea that social pricing challenges sellers (both on a company market and small businesses) to use innovative strategies: whether to alleviate pain, help out in moments of need or sustain business partners – all of which can be done in a flexible and profitable way. It's time to shift from a conflictual logic to a collaborative economy. From mere profit to a profit that takes into account the macrocosmos, systems, customer orientation, a market moving further and further away from fossil fuels.

The concept works for all levels of sales, at state, regional, and local level. This is because, today more than ever, every state, region, city, and even school or institution must face unique challenges. It is time to leave the Age of Fossils behind!

Summary

Sympathetic prices can make the difference to customers in a series of situations.

The application of these "social" prices does not have an immediate positive impact on the timeline of the company's objectives but finds its place in terms of a correct logic of return in the medium and short term.

Both in the B2C and in the B2B sectors, sympathetic pricing can help improve the image and perception of the company, increasing confidence in them. Offering discounts at the right moment is the most effective proof that a company cares about its customers. This will translate into an improvement in the company's image, (re)gaining greater trust from its customers and new, precious brand supporters.

In the long term this, and possible positive media coverage, will also lead to an increase on the baseline.

Companies should therefore look carefully at how to use social pricing.

They might explore the possibility of linking it to a public event such as a concert, or perhaps to a TV program, a sporting event, or a national holiday, or perhaps offer a discount to customers whose football team has been eliminated from a championship. Or it might be linked to evaluations by big data, which measure personal metrics, introducing personalized markdowns to "sweeten" life's bitter moments.

The data generated by consumers could help to guide customers who've had a bad day and reward those who might win little victories.

Let's imagine an exercise app that records results achieved so that, after a user has walked five miles, they're entitled to a markdown on an energy drink, or perhaps – to remain in the field of wellness – an app that gets you a discount on a healthy salad if you're successfully sticking to a diet.

Although some may see this scenario as a *worst case*, an alarming *Big Brother*-company that invades everyone's privacy, marketers really are moving more and more towards targeted and personalized ads and promotions, both in shops and online.

What is more, the application of this pricing policy in B2B might be spread further to more and more strategical fields for our future, perhaps providing bigger discounts to companies who commit to reducing their carbon footprint or improving their eco-compatibility and the social responsibility of their businesses.

CHAPTER 10
PARTICIPATIVE PRICING

"If it has a value, the customers will put a penny in the plate."

Chris Hufford, manager of Radiohead

Case History

It's the beginning of January; the start of the New Year and the sales are coming up. An opportunity for a bargain: but where? As usual, prices, although discounted, are set by the seller. But suppose it were the other way round: if it were the buyer who decided what price to pay?

Everlane, a San Francisco fashion company with retail points in several cities, such as New York and Boston, but mainly active online, regularly proposes sales known as "choose-what-you-pay": it's the customer who chooses how much.

It's YOU, second person plural, who in this case represents on the one hand a move to get closer to what the customer wants, on the other an immediate communication channel, in line with the age of social networks. The formal "sir" or "madam" of the past has (sometimes regretfully) been abolished by the annulment of space – digital, technological; we are all approachable, just a click away.

In any case, leaving aside the (essential) relationships between people, let's get back to the transactions.

The choice ranges through hundreds of articles for both men and women, from cotton T-shirts to colored cashmeres, upmarket leisurewear, and bags.[1] The discounts vary between 20% and 60%.[2]

That said, how does this sort of pricing work?

Basically, it's a matter of knowing how to manage expectations: customers can't choose any price for any item of clothing, but mainly for unsold or surplus goods. Everlane offer buyers three levels of price options on specific promotional days, for example 26 December 2021.

Nevertheless, it's difficult to imagine customers ready to buy a product with a 20% discount when they could have the same one with a 60% markdown: the label passes the ball to the customer, who decides how much to pay.

At retail points Everlane also reveal the real cost of each item with extreme transparency, alongside the price shown in the traditional manner (this allows the full price of an item to be perceived by buyers as well as the relative bargain price). In this way the company reassures their customers, amongst whom is Meghan Markle[3]: the impression given is that of getting a bargain, even when paying the full price outside the sales. Customers are motivated to buy by the "double bargain." From their own point of view, Everlane gets to move unsold goods.

It's a game where everyone wins out.[4]

"Instead of offering a traditional sale, we discovered that Choose Your Price has given us the opportunity to be completely transparent towards our customers about our inventory process and profit margins," maintains the CEO and founder of the firm, Michael Preysman: "By offering our customers three choices we are able to give them a real sense of value for every item and help them make an informed decision."[5]

When customers scroll over the products, pop-up boxes appear explaining that the lowest price is equivalent to the production cost plus delivery and that the average product includes these plus general expenses. Basically the lowest prices do not produce any earnings for Everlane.

Instead, for the highest price, the box explains: "This price helps cover the costs of production, delivery and our team and allows us to invest in growth. Thank you!"

Normal prices are a few dollars over the highest promotional price and around twice as much as the production cost, but in any case a fraction of the markup by most of their competitors at retail points.

Everlane has been applying this approach to monetization regularly since 2015. The company CEO points to the fact that when they initially tested this pricing, 10% of customers chose the medium or highest price and continue to do so, year after year.

The conclusion that can be drawn from this *history* is that people – all of us when we buy something – appreciate transparent prices.[6]

Pay-what-you-want sales thus represent an opportunity to "build" a trusting relationship with buyers.

It pays to communicate transparently the impact of a customer's purchase on your business, on the basis of how much they have chosen to pay.

Buyers will appreciate the fact that they are not only being offered options but also that they are not being made to pay too much for something.

When asked where they got the idea of offering this pricing model, Preysman quotes the online sale of Radiohead's *In Rainbows*, a 2007 album launched by a sort of pay-what-you-want model: "We found the results [of the band's experiment] extremely interesting because few [paying] listeners actually paid the lowest price."

The album's release was indeed a success: of the 1.8 million listeners who downloaded the album, 60% decided to pay nothing and 40% paid – on average – $2.26: "In terms of digital income, we earned more from this album than from all the previous ones put together," declared Thom Yorke, Radiohead's frontman, in an interview.

In addition, once the hard copy came out, sales were not influenced by the digital, pay-what-you-want, advance sale.

The album reached top place in the Billboard classification and sold 3 million copies.[7]

The case in question was a success because Radiohead are . . . Radiohead! Who can forget the video where Johnny Depp and Charlotte Gainsbourg meet in a record store, she looks at him, he pretends not to look at her, they smile at one another, with *Creep* playing in the background, one of the biggest tear-jerkers contemporary music has come up with.

In the end, say Radiohead, you mustn't wait for happiness to come along but grab it, believe in it, even pursue it. Run, even if you don't know where to. But act, for heaven's sake, go out there and get the elevator to happiness!

For this and other reasons – the mood, the sense conveyed of life in various shades of color – Radiohead have fans who are happy to pay for their music, lyrics, and the freedom that can only come from supporting the band and their work.

In a nutshell, it can be concluded that the "pay-what-you-want" model works on the basis of customer loyalty.

Companies with loyal customers can therefore generate reasonable income using this type of monetization.

A study even discovered that buyers are more willing to pay extra if they know the owner of the business.[8]

If, for example, your customer base is particularly devoted to the brand, you might consider using pay-what-you-want pricing as a temporary strategy.

From San Francisco to London: here we find the cashmere fashion producer, London Cashmere Company.

And here too, in a similar way, customers can choose how much to pay. The markdown codes are provided directly – "CWYP15", the acronym of *choose-what-you-pay 15%*, and similarly, the alpha-numerical codes "CWYP25" for the 25% offer and "CWYP35" for the 35%.[9]

Pay-what-you-want pricing can also be adapted to modify the sense of a purchase and make it seem to the customer that they are buying not only a product but also contributing to a certain idea of the world.

On 7 November 2015, for example, the American retail chain 7-Eleven let their customers decide how much to pay for Slurpee drinks, as a donation to organizations fighting world hunger.[10]

Letting customers decide how much to pay may come as a surprise to us if we think of profits, as in the case of Radiohead, but can also reveal how much customers really appreciate a product.

If, instead, customers pay less than hoped, we nonetheless have some fundamental feedback in terms of money: so perhaps the time has come to review marketing strategy.

Analysis of Context

Participative pricing,[11] which groups together concepts which are sometimes synonyms like *name your own price* or *pay-what-you-want, choose-what-to-pay* or *pay-what-you-think-is-fair*, is an approach to monetization where buyers can independently choose the amount to pay the company selling a product or service. We also speak of a free price. The price can start out as completely without charge, or zero euros, or at an extremely low level and rise as desired. At times, a minimum threshold is indicated or a suggested price, to guide the buyer. In the following cases, we shall look at some slight variations on this approach to monetization.

Choose-what-you-pay (or *pay-what-you-want*)

An advanced form of participative pricing is the *choose-what-you-pay* model, also known as "pay-what-you-want." Here the customer pays what they want, sometimes without the supplier being able to decide whether to sell at that price or not, at other times with limits, for example a minimum price.

This type of pricing makes it possible to differentiate prices taking into account consumer diversity and at the same time to allow buyers to exercise some control over the end price of the transaction, thus participating in the process of setting the price.

Participative price-setting, with the greater control perceived by buyers, leads to more intentional buying.

The pay-what-you-want model is increasingly attracting the attention of the market.

The amount of a transaction depends on social preferences in terms of a fair division of value between customer and supplier. Moreover, the idea of keeping the supplier on the market in the long term, also plays an important role.

Using the pay-what-you-want model, the zoo *Allwetterzoo* in Münster, Germany, ran several campaigns which led them to increase the number of visitors *five times over* to 76,000, in less than a month and raise turnover by 2.5 times.[12]

Although every visitor only spent an average €4.76 compared to €10.53 the previous year, the considerable increase in the number of visitors more than compensated for the lower price.

It is perhaps improbable for this type of pricing to sustain a zoo in the long term, all the same, the local police force in Münster –enthusiastic about the success obtained by their animal friends – wishes to consider introducing a model of this type for humans, for paying traffic fines. . .

An initial example of paying what you want in the field of hospitality is the hotel OmHom, an adorable little hotel in the hills above the Cinque Terre in Italy, managed by the entrepreneur Luca Palmero. Starting out from complete transparency of pricing, which aims to help guests have a clear vision of the costs of managing the hotel, a price of $200 a night is suggested, of which 39% goes to the workforce, 20% to suppliers and services, 19% to services such as electricity, 17% to the management and 5% to marketing.[13] Palmero also offers what he calls the "suspended stay."[14] This is an idea borrowed from the Neapolitan *caffè sospeso/*"suspended coffee." In the "unchallenged capital of the *espresso*," where a coffee at the bar is considered a matter of dignity and a fundamental right of every citizen, it is common practice to pay for an extra coffee for the benefit of anyone who can't afford one. In choosing how much to pay, Palmero thus sees a considerable human component.

A second example comes from the IBIS hotel in Singapore, part of the French chain Accor.[15]

The model has also been tested for admission to the Hall of Peace in Münster's historical City Hall. Here there are no more visitors than usual but "the entrance fee paid was slightly above the normal price." Usually entrance to the Hall is possible at a price of $2 for adults and $1.50 for children. We attribute the difference in the two Münster experiments to the different levels of pricing. For the sake of a story, we can report that the same test was run in a cinema, where customers paid far less than the usual prices, as at the zoo.

The manager of the Schmidt theatre on Hamburg's famous Reeperbahn is pursuing a similar plan: audiences only pay what they consider to be reasonable.[16] Even those who pay just $1 get a ticket. The same goes for the Schauspielhaus theatre in Zürich: once a month you pay-what-you-want until seats and tickets are sold out.[17]

A series of cases where choose-what-you-pay is applied are also to be found in the field of catering, wine-bars, hospitality or similar services. After consuming or when paying the check, the customer pays a price they choose. The providers of the service put themselves in the hands of the customer with regard to price. Here, there may be a certain number of customers who do pay a sum that covers costs but, at the same time, there will be others who take advantage of the situation.

Unlike the zoo or the cinema in these cases we are looking at variable costs that increase the risk to the supplier of the service. Here are some examples.[18]

Der Wiener Deewan, a popular Pakistani restaurant in Vienna, serves a buffet to whoever wishes to taste Pakistani food: the choose-what-to-pay buffet offers five different curry dishes, three vegetarian options and two with meat. In addition, every first Monday of the month it's possible to listen to a play-as-you-wish jam session.

In London, following the début of "choose what to pay" at the Little Bay restaurant in 2009, which succeeded in invoicing an extra 20% compared to the fixed-price menu, many other venues, such as the Galvin at Windows, tried the same solution; here the price varied at the customer's discretion, from £25 to £65 per menu.[19]

The *Weinerei*, a wine bar in Berlin, has introduced a choose-what-you-pay policy for wine: after 8 p.m. a glass of wine costs €2; what is more, by paying this "symbolic" sum of money a sort of unlimited access is acquired, allowing the customer to drink all the wine their liver is capable of processing. Before leaving the bar, the customer pays what they think the wine consumed was worth.

In the USA at Jackson, Tennessee, ComeUnity instead offers a menu that changes every day, serving mostly organic, locally sourced, and seasonal food, again using the pay-what-you-want formula. ComeUnity's mission is to love, nurture, and give dignity. If the customer can't pay, ComeUnity offers the opportunity of exchanging an hour of voluntary work for a healthy, hot meal.

There are also chains in a series of venues in the USA, India, Malaysia, Indonesia, Singapore, the United Kingdom, Japan, France, Spain, and Dubai, such as Karma Kitchen: at the end of your meal at any Karma Kitchen restaurant, you get a receipt with $0.00 and a message that reads: "Your meal was a gift from someone who came before you. To keep the chain of gifts alive, we invite you to pay in advance for those who come after you." Guests can pay however they like, in cash or with their time.

Burger King has used pay-what-you-want, but only for the whopper burger and for one day only, giving the income to charity.[20]

In the context of computer games, too, we find *name your own price*: since 2010 Humble Bundle has offered little collections of games that can be purchased at a price set by the buyer.[21] Idem, the *bibisco* software, designed by an Italian for writing novels, follows the choose-what-to-pay model, in which he strongly believes.[22]

Some museums have started to follow the same path: for example, the *red dot* museum in Essen, which devotes a space of 4,000 meters to exhibiting products that have won international design awards, every Friday allows visitors to choose how much to pay.[23]

In consultancy this type of pricing is also present: the Belgian consultants Kalepa, specialists in *experience management* let their customers choose how much to pay for the sessions they call "inspirational," following the philosophy of one of the founders, who actually did her research thesis on the subject of pay-what-you-want.[24]

In 2021 in Scotland, at the height of the pandemic, the first bookshop opened in which you pay-what-you-want. The operation was inspired by a specific philosophy addressing those who detest waste but love books, and therefore believe they should be easily accessible to everyone. In this case the aim is to cover costs rather than generate profits.[25]

Additionally, online, there are virtual bookshops such as Open-Books, which in 2016 had already started allowing readers to pay what they wanted for eBooks, as long as they had read the book first.[26] In this case the moment of "naming the price" shifts from *ex ante* as normally occurs, at the beginning of the transaction and before the consumer experience, to naming the price *ex post* and postponing price setting to a phase that comes after the consumer experience.

The system of choosing what to pay *ex post* works as a signal of quality for attracting readers who don't want to run the risk of buying something that doesn't come up to their expectations.

Even in football we have experienced the "choose-what-you-pay" model.

The Canadian Premier League team Atlético Ottawa applied this type of pricing in 2021. For the first game at the TD Place stadium in Ottawa, 15,000 fans had the opportunity to buy pay-what-you-want tickets. Starting from $0, ticket prices were available at $5 increases, the maximum cost being set at $50 per ticket.

The Atlético Ottawa managers later revealed that the reason for this opening offer was to help mitigate some of the obstacles that members of their community might have faced during the Covid-19 pandemic and allow whoever wished to, to go and support the club in their home games.[27]

At times, a few extra dollars are necessary to get to the end of the month between one wage packet and the next. The only option for those who have trouble obtaining credit from the banks is to borrow money from a friend or family member or to apply for a loan at high interest. A new service called *Activehours* offers an alternative: it gives you access to "your" salary while you're earning it.

This is how it works: the users can obtain an advance on their next salary for the hours they have already worked, up to $100 a day. The most important novelties? No interest and no commission is charged unless the applicant wishes to pay for the service.

Activehours is supported by what is called "voluntary suggestion" by users: "You decide what you want to pay or what you think is fair and you can even decide to pay nothing," declared the founder of Activehours, Ram Palaniappan: "There are some folk who constantly give us a contribution and others who do so every third, fourth or fifth transaction. So we're seeing some very interesting models of contribution."[28]

When you enrol in Activehours you supply the number of your current account.

When you need money, all you have to do is forward a screenshot of a timesheet: you decide how much must be deposited in your account and what contribution to authorize if / in the case that: in a sense it's a classic example of *if / then*.

The app provides five contributions which are suggested for each transaction: 0 is always the first option. For example, on an advance of $100 the contributions suggested would be: 0, $3.84, $5.68, $7.89, and $10.99. As the loans are only generated for a short period of time (for example a week), even a contribution of 1% corresponds to an extremely high rate of interest. Moreover, it can be supposed that the customer wishes to use this credit service repeatedly and therefore pays the "contribution."

One variation of the pay-what-you-want model consists of components with variable prices that basically depend on how satisfied the customer is.

This method is sometimes used in management consultancy. As well as a fixed quota, a variable component can be agreed on, the amount of which is determined by the customer, who evaluates their satisfaction on a predefined scale. In this case, too, the supplier places itself in the customer's hands. When the consultancy agency finds itself faced with the alternative of allowing an unconditional discount or agreeing a price component based on satisfaction, the latter is preferable.

Gratuities are a further variation of the choose-how-much-to-pay model. Normally the customer decides what to pay over and above the price formally asked: this is true for services like catering, haircuts, or portering. Nonetheless, there are also systems in which the contribution is not really voluntary. In American restaurants, for example, you "must" leave between at least 10% and 20% as a gratuity if you want to avoid negative reactions or questions from the staff. These gratuities often constitute the server's biggest remuneration compared to the fixed wage.

Lastly, donations can be interpreted as a third variation on the pay-what-you-want model. In this case, however, it is not strictly correct to speak of "prices" because no tangible or claimable counterpart corresponds to them.

Name your own price

"Name your own price" is a monetization strategy in which sellers allow buyers to decide the end price they wish to pay for the offer, but the transaction only takes place if the offer is equal or superior to the threshold price, which is not revealed by the seller.

The transaction works as follows: sellers list the products with a threshold price beyond which the offer will be accepted. This threshold price is not visible to the buyer. Once the buyer likes the product, they make an initial offer for it.

The sellers' point of view is based on the expectation that the customer will reveal their true willingness to pay a price. And the buyer's offer is binding. Payment is ensured either by providing a credit card number or by direct debit.

If the price offered by the customer is equal or superior to the threshold price, the transaction takes place at the price indicated. If the offer is lower than the threshold price set by all sellers, the buyer has the opportunity to update their offer in the coming rounds.

Reverse auction is also involved here.

In a traditional auction, a seller offers a product or service for which a number of buyers compete. The buyer who can afford the price has the possibility of securing the service or product.

Instead, in a reverse auction, as suggested by the concept, the roles of buyer and seller are reversed, that is, sellers who can provide the service at the price indicated by the buyer win the auction.

The American company Priceline is considered the inventor of the name-your-own-price model, later emulated by firms such as Hotwire.

Starting out from the consideration that airlines regularly flew with only two-thirds of their seats occupied, thus with millions of empty seats, they asked themselves: what would happen if we exploited the Internet to guide demand, filling planes or hotels? Suppose the customers could make an offer – their *own* price – rather than paying the whole price?

Initially the response from the airlines was skeptical: the carriers didn't want to cannibalize prices.

Instead, the founders of Hotwire decided to anticipate their skepticism by using the word "priceline," the line or the price point below which they did not wish to sell, but which would encounter sufficient demand to fill the capacity available.

In this context, "opaque" pricing is also spoken of, since the companies sell their goods at lower, hidden prices.

The target customer is the one who bases their purchase mainly on price: customers choose the route they want to fly or the place and dates and (for hotels) the number of stars.

After paying, the website reveals the flight schedules, airline, and any stopovers, or the name of the hotel, but no reimbursement, modifications, or cancellations are allowed.

According to priceline.com's website: "The *Name Your Own Price* service exploits buyers' flexibility, allowing sellers to accept a lower price for selling their excess capacity without damaging their existing distribution channels or retail price structures."

Today this approach to monetization can be found mainly in the field of music.

This is how Michael Stipe, the famous frontman of the group R.E.M. which broke up in 2011, started to offer tracks to his fans following this approach to pricing: for example, he proposed a package related to the track *Drive to the Ocean*, which relaunched eight elements including the official video, photos, background images for PC and tablets, as well as the lyrics, asking music fans to pay what they wished, starting from 0.77 cents.[29]

Bandcamp, a web service where musicians and groups sell their music to fans allows those who intend downloading the tracks to name their own price when they buy the music; in the same way, the band has the possibility of setting minimum prices for the music produced and the buyers can pay as much extra as they wish.[30]

Economic accessibility is one of the main problems that still affects the clothing industry and this is why Garmentory has set out to create waves in the sector. Garmentory is a virtual market that sells designer goods and items by contemporary boutiques. It also allows customers to suggest their own price.[31] The move has created a new space where new designers can interact directly with customers and customers can obtain a new designer outfit at a price they feel comfortable with.

The well-known Gap brand also followed this approach, though to the letter, to sell items of clothing through a promotional campaign baptized Gap My Price.[32]

Another significant case regards eBay with its "Best Offer," which allows the buyer to offer the seller a price they are "willing" to pay for the object. The seller can accept, refuse, or respond by proposing a different price.[33]

In any case, the name-your-own-price model hasn't had the success initially hoped for: many customers have made unrealistically low offers without generating turnover.

Priceline still exists, but with a different business model. Today the company is part of the market leaders for online travel Booking Holidays, a big player that earned around $7 billion in sales in 2020 (and pre-pandemic over $15 billion in 2019[34]) with a quote of $95 billion on the stock exchange.[35]

The greatest contributor is booking.com, which originated in the Netherlands. The name-your-own-price model makes only a minimum contribution to turnover and has been replaced by Express Deals,[36] that is, offers where Priceline guarantees the lowest market price, otherwise the customer is entitled to a 200% reimbursement of the price difference. Essentially this model is used to trade unsold offers to consumers who are extremely price-sensitive and ready to put up with inconveniences such as repeated flight changes for a particularly low price.

Despite its interesting potential in terms of revealing customers' willingness to pay the prices, the name-your-own-price model has not so far come up to expectations, though this does not rule out a return in the future or its suitability for remarketing.

Summary

Participative pricing is an approach to monetization in which buyers can independently choose the sum they will pay a company selling a product or service. The price can start out from free, or zero dollars, or from a very low level and rise as desired.

At times a minimum price threshold is indicated or a suggested price to guide the buyer.

In the *choose-what-to-pay* model, also known as "pay-what-you-want," companies let consumers decide prices on their own, thus giving up what is traditionally a central managerial prerogative for managing a profitable business.

Non-profit organizations have used this option for a long time in order to attract a vast customer base, but recently many profit-seeking businesses have successfully adopted pay-what-you-want for a vast range of products and services, such as digital books, headsets, music albums, authors' rights, aftermarket support services, restaurants and even business consultancy services.

The recent peak in popularity is due to benefits that extend beyond market penetration, such as short-term promotional prices, prices for reducing piracy, bundling with donations directed to users.

Name your own price is a procedure in which, from the seller's point of view, it is expected that the customer will reveal their true willingness to pay a price. The price offered by the customer is binding. Payment is ensured by providing a credit card number or by direct debit. As soon as the customer's offer rises above a minimum price known to the supplier alone, the customer wins the contract and pays the price offered.

There is a basic difference between the pay-what-you-want and name-your-own-price models. In the latter case, it is the seller who decides whether to accept or refuse the price offered by the customer. In the pay-what-you-want system, consumption and use come before payment. Or else the customer pays what they want in advance, for example at the entrance.

The big question these pricing models answer is, in practice, indirectly rhetorical: why should a customer pay up if they aren't "obliged" to?

Well, many cases and examples demonstrate that people pay even if they are not "obliged" to.

There is a vast field of thought behind this reasoning, an inherent need to be fair. Understanding and applying this logic is thus essential to successfully introduce participative pricing inside your company.

Letting customers decide how much to pay may turn out to be a surprise in terms of earnings, as in the case of Radiohead, but it can also reveal just how much customers really do appreciate the product offered.

If, then, customers pay less than hoped, the message, although negative, will be equally important: perhaps the time has come to review *our* marketing strategy!

CHAPTER 11
NEUROPRICING

"Your brain takes decisions up to ten seconds before you're aware of it."

John-Dylan Haynes

Case History

Every manager's dream: to put prices up and have customers who are happier today, with a higher price, than when they paid less.

Thanks to the use of Neuropricing this dream came true in the Weissenhaeuser Strand tourist village on the Baltic Sea. In this village it is possible to choose accommodation in an apartment or in a hotel, inside a vast leisure facility.

"Many companies miss out on income and profits without realizing it," this is the philosophy of David Depenau, head of the tourist resort.[1] On consideration, the tourist village was offering prices that were too low and (apparently) guests didn't feel at ease with that level of pricing. Especially at peak season, in summer, the prices were perceived by customers as being "too low." Yet, despite this the average was around $200 a day. Even so, too little for a clientèle looking for a high-quality, relaxing holiday, as Depenau now knows.

"Feeling at ease," are words to bear in mind.

Here and now, as we write, the tourist village makes an extra million in turnover and profits, compared to its previous income before prices were reviewed.

Depenau rents out a total of 1,200 apartments. The increase at once translated into higher profits, without losing customers or leaving them dissatisfied. The manager himself finds this an "almost perverse" attitude but is quite ready to talk openly about it.

In the end, he says, everyone is satisfied. He certainly is, but so are his guests. They are far more satisfied now than in the past, Depenau has discovered. A survey shows that they have felt better since the manager raised the prices. What is more, in summer the tourist village is completely sold out.

How was it possible to raise prices and at the same time customer satisfaction?

Depenau indicates that investments for modernizing the village have certainly supported the price increase but that without neuropricing nothing would have made sense, far less so the rise.

And what is neuropricing?

This degree of satisfaction – Aldous Huxley would say – comes because different levels of "perception of reality" are involved. Basically, the world doesn't exist but just our way of looking at it.

And regarding the "right distance" for observation, let's take a closer look at this pricing model.

To carry out a study like this, it's possible to rely on neuromarketing experts like Kai-Markus Müller: Müller's most important tool for discovering how much a customer is really willing to pay lies on the bench in his laboratory.

It's a sort of perforated swimming cap. Once it's on your head, one of Müller's assistants connects up the electrodes. The gel on your scalp makes the measurements more accurate.

It does look a bit like an experiment by a crazy scientist, the one from *Back to the Future* for instance, but then what is science, if not a plethora of experiments, tests, trial and error?

According to Müller, then, in any case people do not do or buy what they say they want.

What customers would really buy and at what price is, however, revealed by their brain waves; in the strict sense of the term, these are not what Müller measures but rather the fluctuations in tension on the upper wall of the scalp, using an EEG-electroencephalogram.

On a screen, it can be seen that the fluctuations in tension move up and down. Every electrode has a codified band width and a color. Thoughts are tone and chrome; to every emotion is associated a shade, a chromatic lemma. It's the matter of seeing the world according to what we are like. What color are we? At every instant that changes (us).

Müller has also tested Starbucks products by putting some simple questions to a sample of customers: for example, how much would you pay for a black coffee in a small paper cup? In such studies, the customers are shown every other second potential prices charged for the product. Based on big data and machine learning algorithms, the EEG reveals for each price whether the price associated with a product is an optimal price. The brain of the test person cannot lie: with EEG it was thus possible to identify optimal prices for Starbucks.

The surprise discovery: Starbucks actually is expensive, but people are willing to pay even more. On the basis of the study, the chain decided to increase the price of a small black coffee from $1.80 to $1.95, without any dip in their turnover being noted.

So what is cost, time, how much does a dollar "cost," what's the value of a break in a bar, with the fan spinning above you, gimme just ten minutes of silence before work, a laugh with friends, a glass of water just before a job interview, the rays of dust floating down from the big central window?

Another exemplary case is that of the Pepsi branch that wanted to know how sales would change if the price of a bag of crisps in Turkey increased by 0.25 Turkish liras.

A market research study of customers showed a 33% drop in sales.

The survey, based on the neuropricing method and run parallel to the traditional one, forecast instead a drop of 9%.

Once the price was raised, the actual drop was just 7%. The neuro-pricing study keeps persons engaged in the price task, asking the question: "Cheap or expensive?" "The longer the interviewees take to reply expensive/cheap, the better it is: the more closely the price indicated corresponds to their perceptions", explains Müller.[2] Based on these insights, Müller developed a scalable tool termed Neuro-Pricing Online, which is being used when numerous products, target groups, markets, or product variations are being tested.

Brain scans lead to the surprising discovery that the best price is often higher than producers or retailers suppose: "The seller's anxiety about price is often more acute than the buyer's," Müller concludes. And the same is true for David Depenau. Both know, as well, that you can't overdo prices. Otherwise sales crash and so do your image (the customer's perception, your reputation) and your profit margins: "It's like standing on the edge of a cliff overlooking the sea," Depenau sums up – the sense of giddiness, the risk of beauty: "If you go a step too far, you fall."

Analysis of Context

Can gut decisions really be taken and even turn out to be "right"? Again. Can your heart really be broken, or can you really follow your instinct?

For the love of poetry, people like to attribute emotions, ideas or actions to certain parts of the body.

Neuroscience or research on the brain demonstrates the opposite.

All decisions, thoughts, and the content of our memory consist in models of activity and interconnections that develop through the over 100 billion neurons and nerve cells all situated within the infinite confines of a single organ – the brain. Where do dreams come from? And suppose the brain were vaster than the sky. . .

Not last in making public the notion that *You are your brain,* was the well-known researcher Manfred Spitzer. And as far as choices linked to pricing are concerned, if the brain of a potential buyer is the decisive organ, for marketing and pricing (in particular) it becomes essential to identify what products generate enthusiasm in the consumer's brain and at what price the buyer's brain registers, or would register, the signal to okay purchase of these products.

All this might seem too mathematical, if it weren't for the fact that at Stanford University in California, Brian Knutson and his colleagues have studied some of these questions with the help of fMRI (functional magnetic resonance imaging) brain scans.

The method makes it possible to determine which areas of the brain are activated according to the input received.

In the brain scanner the volunteers were faced with authentic decisions on purchases.

As a first test, by using a mirror, Knutson *et al.* project the image of a product in the scanner. The more the people being tested liked the product that has just been shown, the more blood is supplied to an area in the brain called the *nucleus accumbens*.

It is interesting to note that activity in the *nucleus accumbens* usually takes place when the brain expects a reward, which is why this part of the brain is considered part of the brain's positive feedback system.

A few seconds later, a price is briefly shown beneath the product.

Lastly the participants in the test had to decide whether to buy the product or not.

If the price was lower than the participant's maximum willingness to pay, activity increased in the brain area of the "medial prefrontal cortex," which is thought to be part of the brain's decision-making system.

If, instead, the price was above the subject's maximum readiness to pay, the medial prefrontal cortex was less active. In this case, scientists measured greater activity in the *insula* brain area, an area usually associated with pain perception, amongst other basic emotions.

Researchers have come to the conclusion that desirable products evoke reward and that high prices lead to sentiments similar to pain, in this case speaking of the *pain of paying*.[3]

What appears to be of some importance to marketing is that Brian Knutson and his colleagues have been able to use the results of the brain scans to predict an actual purchasing decision by the research participants in the brain scanner.

The forecasting ability of the scan data was far better than that of a traditional survey. This discovery can be put into practice. Analytical algorithms of brain scan analysis by electroencephalogram make it possible to measure maximum willingness to pay directly through brain activity.

The advantage of the EEG and fMRI brain scans is that the brain does not lie. The typical problems of classical market research through questionnaires, such as subconscious prejudices, the difficulty of expressing sentiments in words, or deliberate deception, are elegantly sidestepped by measuring the brain directly.

We shall look now at a series of characteristics linked to brain function in relation to prices and their implications for price management.

Low prices do not ruin profit margins alone

Price erosion is harmful to far more than the profit margin "alone." It is a differential karst, a powerful undermining process that has very little to do with "simple" purchasing and a lot more with the "thing" that is being purchased and on the basis of what perception, to gratify what sort of vacuum/trauma/lack or, on the other hand, desire/impetus/happiness (admitting there's anyone who knows the atomic number of the latter).

If traditional surveys are to be credited, what most consumers ask for are low prices. But human psychology is playing a dirty trick on us here (indeed, "God doesn't play dice" goes the famous comment by Albert Einstein to his friend Niel Bohr) and certainly market forces do not comprehend everything.

Consumer associations, which strongly believe in protecting the consumer from businesses presumed to inflate prices, are not always on the side of justice.

The relationship between price and well-being is highly complex. Of course, quality influences price. But it is surprising to find that this well-known effect also applies the other way round. That is, price also influences quality.

Two internationally renowned studies clearly demonstrate the effect of prices: the celebrated behavioral economist Dan Ariely distributed presumed painkillers, mostly simple placebos, to the participants in his study. Half the participants in the test were given a pamphlet explaining that the medicine in question was a recently approved painkiller costing $2.50 a dose. The other half of the subjects were told instead that the drug cost a mere 10 cents.

The participants then received a mild electric shock and their reactions to the presumed drug were recorded: the effect of the more expensive one was judged to be significantly better than the cheaper one, even though there was no difference between them.

The neuroeconomist Hilke Plassmann went even further in another study: the researcher had the participants in her research take part in wine-tasting, analyzing the brain's reactions with the help of a brain scanner.

In Plassmann's test, the same wine was presented as a $10 bottle in one case and as a $90 one in the other.

According to participants, the more expensive wine tasted twice as good. Not only this: the brain scan showed that the areas of the brain associated with positive sentiments were more strongly activated in the case of the more expensive wine than for the one declared to be cheaper.

A question of predisposition? Or of expectations? Or again a sub-conscious prejudice projected onto an object?

As usual, the wisdom of popular sayings holds: "Cheap is cheap" or "What costs nothing is good for nothing."

Low prices are not only harmful to the company's profitability, as Ariely's and Plassmann's results show, but they negatively influence the quality for consumers, too.

Higher prices lead to more profit but also to greater satisfaction and, in the end, to better life quality – this is the conclusion companies should bear in mind when weighing up if and how to modify their prices.

Time and price

The brain is impatient.

In theory we expect human beings to make rational decisions when, for example, they are asked if they prefer a bird in the hand or two in the bush.

On a rational and abstract level it makes sense to opt for what offers most value.

In the same way, we would expect a worker to prefer investing money in pension funds to ensure a dignified old age, rather than spend all their money on consumer products and holidays, thus getting to retirement age with no savings; or a student to prefer two free meals in two weeks' time, rather than one free meal today.

In a piece of empirical research participants were offered the choice of three types of reward: one more limited but immediate, another bigger but further away in time.

The vast majority decided to accept a $10 coupon valid immediately rather than a $100 coupon valid in two months' time.

Even when faced with the choice of an immediate $100 win or $200 in 3 years' time, most of the participants chose the immediate offer.

A short-term reward obviously generates an image of what can be bought immediately, making the option (lower in terms of value but available earlier) far more attractive.

This is why many people prefer to consume today instead of saving money for their old age: from a psychological point of view this has to do with the fear of death, illness, our natural tendency to avoid or escape from the deadly consequences of being alive.

Probably, it is also one of the reasons why a number of insurance policies are cancelled before the long-term benefits materialize or it becomes possible to claim the premium.

From the point of view of neuropricing this implies that promotion linked to immediate advantages for the customer are more likely to prompt acceptance of the price demanded, rather than, for example, accumulating points that can be transformed into markdowns or free products once a sufficient number of points have been totalled.

When payment causes pain

It has been proven that the price of a product is weighed up in the part of the brain that perceives pain. Price and payment might thus both be assimilated in the brain with a negative cerebral impulse. One of the most interesting aspects of this function is that the intensity of the negative sensation does not depend so much on the absolute sum as on the benefit deriving from the purchase.

Let's suppose we have estimated $200,000 for the purchase of an apartment in the process of being built which, however, has yet to be furnished.

More personalization, the builder will tell us, means higher costs, we shall see for ourselves.

If, before we are handed the keys, the price has risen 25% – due to all the extra costs for fitting out the kitchen, living room, bathroom, and bedroom – the pain will be fairly limited if the apartment lives up to our expectations.

Vice versa, let's suppose we order a hamburger from a sandwich bar for just $5, but that after an extenuating wait and growing hunger, at the first bite instant disappointment sets in, that is, the hamburger – euphemistically speaking – doesn't satisfy us at all. The immediate effect in comparative terms will be as follows: the pain caused by this (small) outlay will be far greater than what we felt in relation to the $50,000 extra for furnishings that hadn't been included in the estimate, but which is linked to our complete satisfaction when we sit down with a cold beer and some cool jazz and look around at the fine apartment we've managed to put together, wrapped in silence, with the lights of the city far below us. We smile to ourselves – "Cheers!"

Pain thus comes from the sensation of loss rather than the sum of factors involved.

The reason for this is that brain mechanisms balance positive and negative emotions. So the joy of entering a new apartment furnished to our own taste dominates over other emotions: everything seems "positive," we shall see the glass half full with the furnishings we bought on offer, thanks to the markdowns accessible via the builders, for example those in the bathroom or kitchen.

In the case of the hamburger, instead, because of our hunger our expectations were high and, moreover, related to a primal need: we were already anticipating the taste of a succulent burger and the wait seemed to strengthen our expectations of a feast for the palate. The $5 price also seemed to be quite "fair." And instead – after the first mouthful, the anticipation deconstructs and the disappointment – that "Ugh!" attached to the mouthful – causes the positive emotion to change into its exact and decidedly negative opposite, as the brain's pain center is activated and reinforced by thoughts like "Did I really pay for this trash?" Not an existential question but an eminently day-to-day one.

What is more, the memory of this bad experience is not going to disappear any time soon and that sandwich bar will be avoided forever.

The "pain" of payment is thus all the stronger when the price is perceived as unfair.

Returning to the case of the new apartment, a month after moving in, you're now used to the new kitchen and super bathroom as well as the pleasant furnishings, and everything will be summed up as a normal, pleasant habitat, the habitat where we live, eat (not hamburgers), love, smile, and invite our dearest friends. The initial enthusiasm is over, it's true, but so is the memory of the extra outlay.

The waiter at the sandwich bar, for his part, might have noticed the customer's disgusted expression, unless their meals are delivered to their home. After the first bite, if he'd noticed the customer's disappointment, he could have taken action to modify the negative and what's more lasting perception.

If he had promptly offered a different hamburger free of charge or perhaps some other food, he would have substantially reduced the pain perceived by the customer. The client would have perceived excellent service and would probably have remained a loyal customer. These are all "ifs" and if you think carefully about it, all our days are like this: the fruit of crossroads, left or right, I notice the disappointment, take action and correct it; I don't notice and that perception will stick forever in the memory. Inception, semantic fields, and sliding doors. The mind is an (our) inner world that inhabits the surroundings.

It is up to us to fill what we define as *outside*, letting necessities emerge and then acting on that limit, that wonderful and unique frontier-territory that binds us to the world.

A wide range of offer inhibits buying

Amongst (too many) conflicting alternatives, the *non-choice* prevails: our brain halts the buying instinct when the choice is too vast.

This was demonstrated by an experiment in a Californian supermarket[4] in which customers were presented with 24 different types of Wilkin & Sons jams.

The objective was to observe buying behavior and see whether, after having tasted a variety of jams, the customers would buy the one they considered "tastiest."

Interest in tasting the different jams was not long in coming and a number of customers stopped to taste in a special corner of the supermarket. Nonetheless, of all those who had tasted the jams, only a limited 3% decided to buy some.

When the same experiment was run with only six different types of jam, the result was clear: although the goods displayed were less attractive, around one-third of the tasters decided to go ahead and buy. A considerable increase: from 3 to 33%!

One explanation is to be found in the customer's subconscious. Faced with a wide choice, they experience a complexity that increases their perception of the risk of making the wrong choice, so that they withdraw from the purchase.

If, instead, the choice is reduced, the decision appears less complex and thus not so risky in terms of making a wrong choice, so that customers' defenses are lowered in favor of purchasing.

The context influences price perception

Context has a considerable impact both on the perception of price and on willingness to pay. It is quite easily observed that on holiday or out shopping at the weekend money is spent (by the majority of shoppers) far more freely than when doing the weekly shop for everyday needs.

We adapt to the "exceptional" context we find ourselves in and in these cases we buy without paying excessive attention to prices. So we find consumers who on their weekly supermarket shop compare the prices of the different types of pasta on display, perhaps choosing the store's own brand and saving 10 or 20 cents, whilst the same careful shoppers on holiday or at the restaurant with their 'other half' will order a special and fairly expensive wine without even checking the price of it.

The skilled salesperson's task is therefore to set up the context of the buying process in order to increase the potential customer's willingness to pay and relegate price perception to the background, independently of what is being sold where.

Gut decisions take place in the head

With the spread of tablets and "intelligent" telephones – smartphones – more and more information is at hand. But in our case, this complicates choices. In order to evaluate prices, both in a professional context and in a private one, there are some heuristic rules that can be applied, that is, methods for simplifying the decision-making process, which can save us time.

Two important methodologies can be distinguished.

The first is the *heuristics of recognition*: in the choice between two elements, whether products or services, the choice is always oriented towards what is *re*-cognized, what appears "similar" to something we have some familiarity with.

The second is the *heuristics of willingness* and is based instead on willingness or the frequency of cases that are called to mind through experience. It is the equivalent of an intuitive statistical inference but using as samples the memories we have in store from our experience.

The essence of the heuristics of recognition is to find something "familiar," a sort of Proustian *madeleine*. It is of no use to think specifically of determined attributes of the new products or be aware of them, but will suffice for us to be familiar with another generic product or the brand that launched it onto the market, in order to decide to buy it. So, for example, if the consumer is familiar with Ferrero chocolates, they will tend to buy a new product from this producer, rather than a new one from their competitors, Lindt.

The mechanism of the heuristics of willingness instead is based on evidence guiding the purchase: by looking at what most other consumers buy, there is already a guideline there that in some way is understood by the buyer as a suggestion to stick to: at times it's enough to follow another buyer who may seem particularly shrewd or for some reason "on the ball," see what they put into their supermarket trolley and choose the same product. Or you might tend towards the product that a friend, a relative, or a figure of authority, or else – as the new frontiers of social marketing tell us – an influencer, might have bought.

Companies that intend to profit from maximum willingness to pay have an advantage if they pick up on their customers' *heuristic* choices: like this they can direct information and offers in such a way as to promote purchase of their products.

Summary

The neurosciences – in particular neuropricing – help understanding new and future human purchasing models, whether conscious or subconscious.

Neuropricing supports companies to decode the subconscious reasons underlying customers' purchasing decisions.

The corresponding price strategies are optimized in order to increase sales, improve acceptance or communicate the quality of a product.

Algorithms of analyses from brain scans using EEGs make it possible to measure maximum willingness to pay directly from brain activity. The advantage of certain techniques, such as EEG and fMRI, lies in the fact that, unlike us, our brains cannot lie.

The typical problems of classical market research by means of questionnaires – such as subconscious prejudices, difficulty in expressing sentiments or self-deceit even more than self-misunderstanding, or building references on the basis of information – can be sidestepped by direct measurement of brain processes. Bearing in mind a series of characteristics linked to the way the brain works in relation to prices and the implications these have for price management, companies can thus optimize their pricing and guide the choices of their present and future customers.

PART III
HOW TO WIN

CHAPTER 12
SUCCESS WITH NEW PRICING MODELS

"Customers don't buy products but the value they perceive."

Peter Drucker

Case History

You're part of a healthy company that's been growing for 30 years. Turnover: $4 billion. Profit margin, almost 20%. You have an internationally acknowledged and appreciated range of products. Precisely the ambition and objective of most entrepreneurs and managers. A dream.

And then. One day you wake up feeling as though you're trapped in a nightmare: top management announces that the revenue model is going to be completely upturned. As from tomorrow, products will not be sold as they have been in the past. As from tomorrow a new model of monetization, unknown to most people, is being introduced, placing at risk all the certainties acquired over thirty years of hard work and sacrifice.

But – *life is what happens while you're busy making other plans* – the unexpected happens: once the revenue model has been turned upside down, the company starts off again and grows even more, and more profitably, than before!

All this actually happened to Adobe in San Jose, California. When you think of success stories in the field of subscriptions, the so-called "software as a service (SaaS) system," the first companies that spring to mind are LinkedIn, SalesForce, Zendesk. Not Adobe. Wrongly so: in fact it is this software company famous for products like Photoshop, PostScript and Acrobat, that is one of the most successful pioneers on this front.

The year 2013 marked their transition from a sales model based on product to a subscription model.

Adobe traditionally sold their design- and publication-ware in the form of physical products, packed and distributed with a perpetual license, where customers paid once only and then used the software for an undetermined period of time. The model was profitable, too, and Adobe earned a net profit margin of 19%. But this inflexible business model also had some disadvantages.

It did not allow the company to establish a permanent relationship with customers, nor did it allow for updating the software. This is why it stopped Adobe from offering a constant flow of innovations and improvements and with them the possibility of generating a constant flow of income.

The solution came in the form of a radical shift to *Adobe Creative Cloud*, a subscription model based on "cloud" services replacing the old model of discs and licenses: by using the cloud, customers receive frequent software updates, as well as a series of new services online. The approach to monetization changed from an *una tantum* purchase of $1,800, to $50 a month for the entire Creative Cloud (or $19 a month for a single app).

Sustained by a massive communications campaign, the change proved an enormous success. Adobe's market capitalization amounted to $22.5 billion when they launched the new model in 2013. By 2021 it had risen to over $269 billion with an annual turnover of $16 billion.

How to Succeed With New Revenue Models

To deal with changes, timing needs to be anticipated and we ourselves have to become part of the change. The *fil rouge* of this book, the question we mean to answer, is: how can our own revenue models and organization be transformed, so as to obtain an advantage (all the more so in a period that is already in the midst of radical social and economic transition)?

Let's try to create some order in all this, summing up what we have looked at so far and concentrating on just a few, clear passages that explain the coming *Pricing Model Revolution*.

Succeeding with new approaches to monetization means getting clear answers to three key questions:

1. *What value does my customer perceive?*

   Once the customer's needs have been identified – and this is always the best point to start out from – understanding the sources of a product's value helps to draw conclusions as to what can be monetized, from the point of view of a revenue model, as well.

2. *How should the approach to monetization be set up?*

   The new way of monetization must be defined (the practical examples dealt with can be studied, so as to look at similar strengths or weaknesses, as well as specific characteristics, in your own company), with the aim of fully grasping the perceived value.

3. *How is the change in the company's revenue model initiated?*

Moving from concepts to facts: doubts and resistance have to be overcome and "having always sold in a certain way" must change into the drive to create an urgency for change.

Let's look now at a specific analysis of the individual engines of change.

1. Identifying perceived value

What *value* does my customer perceive? This question should always be the starting point for any reasoning about monetization.

We take it for granted that any company that intends to prosper is able to generate value: unless they perceived value, customers would not be willing to pay. On the other hand, the company must be able to reap this "value" through innovative pricing models.

As we have seen, paying for washing machines, engines, medicines, music albums or theater performances is a thing of the past in many contexts.

The model based on exchange of property is anything but optimal and does not make it possible to reap the full value supplied to customers.

Today we know that the "real" need is shiny clean dishes, not the ownership of the tool that makes this possible. Just as the "real" need is the hours of flight that take me to a determined place, certainly not owning the engine of the plane that takes me to my destination!

The technological progress described in the first chapter has sparked off radical change in ways of monetization. For companies this means a combination of possibilities for new strategic priorities:

understanding how the customer uses our products and verifying their real performance.

Thanks to modern technology it is possible to trace the use of products, become aware of the context and applications they are used in and thus lay the bases for quantifying the solution offered and the value linked to it.

2. Setting up the new monetization

How should the new monetization be set up?

Once the perceived value has been identified and with it elements such as the way the customer uses the product or when it is used, we pass on to the definition of the new model of monetization in order to fully realize its perceived value.

The objective must be to acquire customers, getting rid of all purchasing barriers. Today, ownership – the hub of past revenue models – is considered by many customers as the main barrier to buying: the cost for gaining ownership of a product may be considered excessive or there may be the fear that this cost may not be proportional to the use made of it.

But companies have a range of options at hand for overcoming these obstacles and encouraging customers to make a purchase. In the previous chapters we have looked at the 10 main options for offering innovative pricing models. Amongst the approaches to monetization we have seen that some models are based on price sharing, on unpacking or on enjoying use of a service, extending in this way the customer base that accesses our products and/or services (aware that in any case with the traditional "ownership" model, customers would not have gone on to buy).

By tracing usage, companies break down the product's elements of value, at the same time making the product more easily digestible (and affordable) to customers. This happens, for example, by providing an access platform for several customers or when the seller makes the whole product available to customers, asking them to pay for usage only.

3. Changing the revenue model

How do we go about changing the revenue model in a company? To gain success in today's competitive context, one basic question must be asked: is my approach to monetization still adequate to ensure profits and growth in my company or is a new revenue model needed?

The starting point for finding the answer to this question is an analysis of the existing type of revenue model, to make sure it is capable of making the most of the customers' willingness to pay.

Successful companies analyze various options of revenue models, sometimes also testing new ways of monetization, to see whether and in what way the pricing applied and the company's products or services sold need to evolve.

At times these evaluations can lead to the coexistence of several models which in the short term may cause conflicts inside the organization, whilst in the medium term they may then make it possible to provide a better guide to several customer segments, also creating competitive advantages.

This is what happened to HP, where the revenue model evolved with the launch of a subscription offer called Instant Ink.[1] In the past, ink for the printers could only be accessed by a transactional model which involved ownership passing from HP to the customer with all the problems linked to the difficulty of anticipating the moment when the ink ran out and the search for a new cartridge began. With Instant Ink, instead, an automatic delivery service can be accessed, which delivers the cartridges straight to your address.

HP monitors ink levels when the printer is active and connected to the web: the "intelligent" printer thus identifies the ink levels and automatically orders the new cartridges from HP before the ink runs out completely. The risk of being left without ink with all the resulting bother is solved once and for all. Once you have registered with the service, there will no longer be any need to buy the refill cartridges from retailers.

These are small, everyday things, yet important in terms of our most precious possession – time!

And this is the value offered to the customer by a revenue model no longer tied to the sale of the cartridge but to the number of pages printed over a certain time span.

After its 2013 launch, by 2022 the number of subscribers had risen to over 10 million users: a clear success for the new revenue model, which runs alongside the traditional one where cartridges are purchased.

On the one hand HP has managed to give its customers more choice, making it possible – thanks to the new revenue model – to win new market shares by satisfying latent demand. On the other hand, in the short term, HP has probably cannibalized part of its cartridge sales, creating new tensions both in its own sales force and in retailers, who are no longer the intermediaries of direct sales.

Decisions about the form of a revenue model depend on the value created for customers, on the competitive context and on the speed at which an organization is capable of evolving towards the new way of generating profits. The main key to success in this sense is how far our organization manages to overcome resistance to change.

Lessons on changing the form of monetization

A lot is to be learned from the case of Adobe, on how to successfully introduce a new revenue model and revolutionize a company's offer.

Nonetheless, this transition did not take place from one day to the next: Adobe had almost 30 years of experience, customers to keep happy, and users to persuade. Consequently, the revolution was carefully prepared by means of efficient rethinking of the business model. The company then launched the process that finally allowed them to become a clear leader in subscription economy.

Here are six lessons that may be drawn from that shift in revenue model.

1. Formulate a clear vision with tangible objectives

Adobe's top management had a clear vision: the traditional ownership-based model would become a burden to growth. In the future it was the customer's access to the product's innovations that was to be preferred. Adobe therefore created a new series of metrics for their subscription service, provided instructions for the interested parties and kept their promises. Their objective included 4 million subscribers by 2015 and an increase in recurrent annual income. According to their financial director, Mark Garrett, these milestones captured the interest of investors in the company's long-term objectives.[2] In turn, they contributed to making it clear that SaaS was Adobe's future.

2. Persevere along your path

Adobe's transition to the new revenue model was not welcomed with enthusiasm. On the contrary: 30,000 Adobe customers signed a petition on Change.org asking Adobe to abandon the shift towards SaaS. This was an extreme reaction, in view of the fact that Change. org is a platform for petitions regarding social causes. Top management, however, remained firmly convinced that the shift to SaaS would allow them to supply a better product, more easily updated, faster, and more secure, as well as being one that could regularly be improved. The SaaS model was also established as a way of extending the customer base and, consequently, cashflow. And so, Adobe took their decision and stuck to it right to the end. They did not desist or stop when faced with obstacles. A clear example of the strength of your own convictions.

3. Do not force the transition or take customers by surprise

Adobe's *SaaS Adobe Creative Cloud* was originally launched in April. Their first subscription version offered customers their services, alongside the traditional software for purchase, an option that was not withdrawn until 2017.

During the time that the offers co-existed, for 5 years the subscription service was available in several versions, before becoming the only option for users. The company acted in such a way as to ensure that nothing about the transition came as a surprise. They announced their intentions to their stakeholders as early as November 2011. Adobe then started to prepare users for the withdrawal of their (now) old Creative Suite shortly afterwards, formally announcing in May 2013 that they would no longer be developing the line of Creative Suite products (although they would continue supporting it).

4. Communicate proactively both with shareholders and with users

At the start of the transition to the subscription model, Adobe published an open letter to users, opening up a dialogue on the coming changes. Their leadership knew that without buy-in from their existing and loyal customers, it would not be possible to shift efficiently to the new subscription-based model. As a publicly quoted company, they also acknowledged that as well as their customer base, their stakeholders would need a detailed explanation and ongoing communication throughout the process.

5. Consider every aspect of the change and prepare to adapt continually

Adobe considered the new service as a completely innovative product, or in other words, as the authentic "digital experience" of its products.

According to Garrett: "Shifting onto the cloud influenced the way in which we designed products, operations, go-to-market and business models."[3] Adobe saw their products and connected functions as a real life cycle (including marketing processes, analyses, advertising, and trading). In other words, Adobe ceased to cling to the *status quo* of trying to change "as little as possible." On the contrary, they saw the shift to SaaS as a way of reinventing and reintroducing products and offers.

As Creative Cloud was destined to provide services for a range of customers, from individuals to big companies, at the start freelancers and amateurs in particular were dissatisfied with the price structure: "Adobe is ripping off small businesses, freelancers and the average consumer. They don't seem to realize that not every company is a multi-million-dollar multinational with infinite resources," said the petition.

The reaction did not go unacknowledged. The company listened to the complaints made by this sub-group of users and in response introduced a more economic version for photography only, which included variations on Photoshop and Lightroom. In this case, too, by not opposing but listening and relating to users, the move proved enormously popular. To sum up, Adobe regarded their transition to SaaS as a large-scale transformation of the company. And, like most successful company makeovers, this took time, incorporated feedback from stakeholders and was achieved through constant progression towards the *new* objectives.

6. Continue to create value

Adobe accepted their customers' challenges and transformed them into opportunities for added value. According to Garrett: "For any company that moves towards a new revenue model, continuous value must be provided to the customer, and new and previously inexistent sources of value must be created that did not exist in the old model. You can't just sell the old offer in a different way."[4] Adobe's cloud products were able to attract new consumers and also maintain many of their existing customers.

Final Summary

A suitable pricing model is one of the most important elements for ensuring business success. If satisfactorily set up, it will allow the company to prosper. If, instead, it is not well managed, it can even lead to the whole company going bankrupt.

In monetization, excellence goes well beyond the optimal management of the individual prices of products in a portfolio. Proper monetization implies the alignment of strategy, objectives, positioning, and also governance, tools and all the processes that regard company culture (in the end expressed in the revenue model which, in turn, translates into prices).

To obtain success with revenue models it is therefore essential to answer these three key questions:

1. *What value does my customer perceive?*

2. *How should the approach to monetization be set up?*

3. *How is the change in revenue model to be initiated in the company?*

To be successful in changing its monetization model, a company should treasure these six lessons regarding a shift in revenue model:

1. Formulate a clear vision with tangible objectives;

2. Persevere on your path;

3. Do not force the transition or take customers by surprise;

4. Communicate proactively both with shareholders and with users;

5. Consider every aspect of the change and prepare to adapt continually;

6. Continue to create value.

Evolution of the revenue model, which we have defined as the *Pricing Model Revolution*, is one of the key challenges of the near future. How about you? Are you ready for the change?

NOTES

Chapter 1

1. Source: Horváth (2022). https://www.horvath-partners.com/en/?hcc= en-us

2. Source: Horváth (2022). https://www.horvath-partners.com/en/?hcc= en-us

3. Source: Horváth (2022). https://www.horvath-partners.com/en/?hcc= en-us

Chapter 2

1. Pay per wash by Winterhalter (2022). https://www.pay-per-wash.biz/ uk_en/ (accessed January 3, 2022).

2. https://www.kaercher.com/de/professional/digitale-loesungen.html (accessed February 11, 2022).

3. Constine, J. (2017). Gym-as-you-go. https://techcrunch.com/2017/ 12/03/gym-as-you-go/ (accessed October 10, 2021).

4. O'Malley, K. (2020). 16 of the best pay-as-you-go gyms, perfect for exercise commitment-phobes. *Elle*, September 30. https://www.elle .com/uk/life-and-culture/culture/a31007/best-pay-as-you-go-gym/

5. Metromile website (2022): "With Metromile, your rate is based on your actual driving habits. Our customers save 47% on average compared to what they were paying their previous auto insurer". https:// www.metromile.com/ (accessed January 3, 2022).

6. Michelin website (2022). https://business.michelinman.com/freight-transportation/freight-transportation-services/michelin-fleet-solutions (accessed January 3, 2022).

7. Michelin website (2022): "Michelin has long been selling kilometers, landings and tons carried rather than tires. These solutions all rely on embedded pressure-monitoring systems that are used to optimize preventive maintenance and minimize vehicle down-time. The advantage of this business model is that customers pay only for what they use, which makes mobility more accessible and more efficient." https://www.michelin.com/en/sustainable-development-mobility/working-towards-sustainable-mobility/more-accessible-mobility/ (accessed January 3, 2022).

8. Rolls Royce website (2022). https://www.rolls-royce.com/media/our-stories/discover/2017/totalcare.aspx (accessed February 11, 2022).

9. Zipcar website (2022). https://www.zipcar.com/en-gb/daily-hourly-car-hire (accessed February 11, 2022).

10. Atlas Copco website (2022). https://www.atlascopco.com/content/dam/atlas-copco/compressor-technique/compressor-technique-service/documents/2935%200173%2020_airplalan_leaflet_en_lr.pdf (accessed February 11, 2022).

11. Visnjic, I. and Leten, R. (2021). Atlas Copco: From selling compressors to providing compressed air as a service. ESADE working paper 275, February 16.

12. https://samoa.un.org/en/130272-food-aplenty-poor-nutrition-undermines-good-health-samoans

Chapter 3

1. Heidelberg website (2022). https://wwtw.heidelberg.com/global/en/services_and_consumables/print_site_contracts_1/subscription_agreements/subscription_1.jsp (accessed January 1, 2022).

NOTES

2. Müller, E. (2019). Die Netflix industrie, *manager magazin*, July, pp. 94–97.

3. Gartner website (2022). https://www.gartner.com/en/newsroom/press-releases/2021-04-21-gartner-forecasts-worldwide-public-cloud-end-user-spending-to-grow-23-percent-in-2021 (accessed January 3, 2022).

4. Justfab website (2022). https://www.justfab.com/how-it-works (accessed January 2, 2022).

5. For an overview of subscriptions regarding pets, see https://hellosubscription.com/best-fresh-dog-food-subscription-boxes/

6. Interview conducted by the author with the Zenises CEO, Haarjeev Kandhari, in January 2022.

7. Horváth Research, January 2022.

8. Horváth Research, January 2022.

9. Barilla website (2022). https://smart.cucinabarilla.it/pages/come-funziona (accessed January 1, 2022).

10. Today the program has been renamed *Porsche Drive* and the approach has also been changed, https://customer.drive.porsche.com/germany/en

11. Horváth Research, January 2022.

12. BMW website (2022). https://www.bmwnews.it/bmw-intelligent-personal-assistant-hey-bmw (accessed January 4, 2022).

13. https://www.bloomberg.com/press-releases/2019-01-22/mann-hummel-partners-with-sierra-wireless-to-provide-predictive-maintenance-in-industrial-and-agricultural-vehicle-fleets

14. Vissmann website (2022). https://www.viessmann.de/de/wohngebae-ude/viessmann-waerme.html (accessed January 4, 2022).

Chapter 4

1. Christensen, M.C. (2006). What customers want from your products. *Harvard Business Review*, January 16.

2. Logan, B. (2014). Pay-per-laugh: The comedy club that charges punters having fun. *The Guardian*, October 14.

3. Google Ads (2022). https://ads.google.com/home/#:~:text=Grow%20 your%20business%20with%20Google,or%20calls%20to%20your%20 business (accessed February 20, 2022).

4. Google Ads (2022). https://support.google.com/google-ads/answer/ 7528254?hl=en (accessed February 11, 2022).

5. Alphabet Inc. (2020). *Annual Report*, pp. 33-34, 66.

6. Enercon (2021). www.enercom.de/en/home (accessed October 20, 2021).

7. Atlas of the Future (2022). https://atlasofthefuture.org/project/pay-per-lux/ (accessed January 2, 2022).

8. Roche (2022). https://www.roche.com/dam/jcr:58422653-e739-47c1-9ba8-e5a29c31de51/en/innovative_pricing_solutions.pdf (accessed January 3, 2022).

9. Bryant, M. (2018). GE, Medtronic among those linking with hospitals for value-based care, *Healthcare Dive*, March 29.

10. Bryant, M. (2018). GE, Medtronic among those linking with hospitals for value-based care, *Healthcare Dive*, March 29.

Chapter 5

1. Freely translated from the *Treccani* dictionary.

2. Thaler, Richard (1983). Transaction utility theory. In: *Advances in Consumer Research*, Vol. 10 (ed. Richard P. Bagozzi and Alice M. Tybout), 229–232. Ann Arbor, MI: Association for Consumer Research. Kahneman, Daniel and Tversky, Amos (1979). Prospect theory: An analysis of decision under risk. *Econometrica* 47: 263–291. Thaler,

Richard (1982). Using mental accounting in a theory of purchasing behavior. Cornell University, Graduate School of Business and Public Administration working paper.

3. Ariely, Dan (2010). *Predictably Irrational.* New York: Harper Collins.

4. Ariely, Dan (2010). *Predictably Irrational.* New York: Harper Collins. Courtesy of Horváth.

5. 99 endings are typically linked to bargain products. More exclusive or premium products would follow a different, non-discounter positioning approach.

6. Waber, Rebecca L., Shiv, Baba, Carmon, Ziv and Ariely, Dan (2008). Commercial features of placebo and therapeutical efficacy. *Journal of the American Medical Association* 299: 1016–1017.

7. Gabler, Colin B. and Reynolds, Kristy E. (2013). Buy now or buy later: The effect of scarcity and discounts on purchase decisions. *Journal of Marketing Theory & Practice* 21 (4): 441–456.

8. Lowry, James R., Charles, Thomas A. and Lane, Judy A. (2005). A comparison of perceived value between a percentage markdown and a monetary markdown. *Marketing Management* 15 (1): 140–148.

9. Tversky, Amos and Kahneman, Daniel (1981). The framing of decisions and the psychology of choice. *Science* 211 (4481): 453.

10. Wathieu, Luc, Muthukrishnan, A.V. and Bronnenberg, Bart J. (2004). The asymmetric effect of discount retraction on subsequent choice. *Journal of Consumer Research* 31 (3): 652–657.

11. Monroe, K.B. and Lee, A.Y. (1999). Remembering versus knowing: Issues in buyers' processing of price information. *Journal of the Academy of Marketing Science* 27 (2): 207–225.

12. Coulter, K.S. and Coulter, R.A. (2005). Size does matter: The effects of magnitude representation congruency on price perceptions and purchase likelihood. *Journal of Consumer Psychology* 15 (1): 64–76.

13. Chaiken, S. (1980). Heuristic versus systematic information processing and the use of source versus message cues in persuasion. *Journal of Personality and Social Psychology* 39 (5): 752–766.

14. Puccinelli, N.M., Chandrashekaran, R., Grewal, D. and Suri, R. (2013). Are men seduced by red? The effect of red versus black prices on price perceptions. *Journal of Retailing* 89 (2): 115–125.

15. Meyers-Levy, J. and Maheswaran, D. (1991). Exploring differences in males' and females' processing strategies. *Journal of Consumer Research* 18 (1): 63–70.

16. Anderson, E.T. and Simester, D.I. (2003). Effects of $9 price endings on retail sales: Evidence from field experiments. *Quantitative Marketing and Economics* 1 (1): 93–110.

17. Inman, J.J., McAlister, L. and Hoyer, W.D. (1990). Promotion signal: Proxy for a price cut? *Journal of Consumer Research* 17 (1): 74–81.

Chapter 6

1. Dahlenberg, A. (2014). Travis Kalanick's take-no-prisoners startup strategy in 9 quotes. *The Business Journals*, November 5. https://www .bizjournals.com/bizjournals/news/2014/11/05/travis-kalanicks-take-no-prisoners-startup.html

2. Shoemaker, S. (2010). Price customization. In: *International Encyclopedia of Hospitality Management* (ed. A. Pizam), 2nd edn, p. 511. Oxford: Elsevier.

3. Metha, N., Detroja, P. and Agashe, A. (2018). Amazon changes prices on its products about every 10 minutes – here's how and why they do it. *Business Insider*, August 10. https://www.businessinsider.com/amazon-price-changes-2018-8

4. https://www.focus.de/finanzen/news/studie-zeigt-preisschwankungen-bei-amazon-um-bis-zu-240-prozent_id_4503019.html

5. https://www.finanzen.net/nachricht/geld-karriere-lifestyle/dynamic-pricing-gleiches-produkt-unterschiedliche-preise-bei-amazon-co-von-dynamic-pricing-profitieren-9153753

6. https://www.businessinsider.com/amazon-price-changes-2018-8

7. Uber themselves explain quite openly: "Dynamic pricing helps us to make sure there are always enough drivers to handle all our ride requests, so you can get a ride quickly and easily – whether you and friends take the trip or sit out the surge is up to you". https://help.uber.com/riders/article/why-are-prices-higher-than-normal---?nodeId=34212e8b-d69a-4d8a-a923-095d3075b487

8. Bhuiyan, J. (2015). Uber is laying the groundwork for perpetual rides in San Francisco. *BuzzFeed News*, August 24. https://www.buzzfeednews.com/article/johanabhuiyan/uber-is-laying-the-groundwork-for-perpetual-rides-in-san-fra

9. Uber Pool: https://www.uber.com/gb/en/ride/uberpool/

10. Uber cities: https://www.uber.com/global/en/cities/

11. Cross, R.G. (1997). *Revenue Management: Hardcore Tactics for Market Domination*. New York: Broadway Books.

12. Dütschke, Elisabeth, and Paetz, Alexandra-Gwyn (2013). Dynamic electricity pricing – which programs do consumers prefer? *Energy Policy* 59: 226–234.

13. https://www.businessinsider.com/amazon-price-changes-2018-8

14. Horváth Research (2022).

15. Richards, Timothy J., Liaukonyte, Jura and Streletskaya, Nadia A. (2016). Personalized pricing and price fairness. *International Journal of Industrial Organization* 44: 138–153.

16. Amaldoss, Wilfred, and Chuan He (2019). The charm of behavior-based pricing: Effects of product valuation, reference dependence, and switching cost. *Journal of Marketing Research* 56 (5): 767–790.

17. Chen, Yuxin, and Zhang, Z.J. (2009). Dynamic targeted pricing with strategic consumers. *International Journal of Industrial Organization* 27 (1): 43–50.

18. Feinberg, Fred M., Krishna, Aradhna and Zhang, Z.J. (2002). Do we care what others get? A behaviorist approach to targeted promotions. *Journal of Marketing Research* 39 (3): 277–291.

19. Bradlow, Eric T., Gangwar, Manish, Kopalle, Praveen K. and Voleti, Sudhir (2017). The role of big data and predictive analytics in retailing. *Journal of Retailing* 93 (1): 79–95.

20. Amaldoss, Wilfred, and Chuan He (2019). The charm of behavior-based pricing: Effects of product valuation, reference dependence, and switching cost. *Journal of Marketing Research* 56 (5): 767–790.

21. Cheng, Hsing K. and Dogan, Kutsal (2008). Customer-centric marketing with Internet coupons. *Decision Support Systems* 44 (3): 606–620.

22. Liu, Yunchuan and Zhang, Z.J. (2006). The benefits of personalized pricing in a channel. *Marketing Science* 25 (1): 97–105.

23. Cheng, Hsing K. and Dogan, Kutsal (2008). Customer-centric marketing with Internet coupons. *Decision Support Systems* 44 (3): 606–620.

24. Conrad corporate website (2022). https://www.conrad.de/de/p/lenovo-thinkcentre-m93p-10a8-desktop-pc-refurbished-sehr-gut-intel-core-i5-4570-8-gb-500-gb-hdd-intel-hd-graphics-1889505 .html (accessed January 1, 2022).

25. Sahay, Arvind (2007). How to reap higher profits with dynamic pricing. *MIT Sloan Management Review* 48 (4): 53–60.

26. https://sloanreview.mit.edu/article/how-to-reap-higher-profits-with-dynamic-pricing/.

27. Zatta, Danilo (2016). *Revenue Management in Manufacturing.* Springer.

28. https://www.michaeleisen.org/blog/?p=358

Chapter 7

1. Orsay website: https://world.orsay.com/de-de/company-information/aboutUs/about-us-company-main/

2. https://blueyonder.com/knowledge-center/collateral/orsay-case-study

3. Die Preismaschine; Künstliche Intelligenz gegen menschliches Bauchgefühl: Wie Daten helfen, höhere Margen zu erzielen, Penner abzuschleusen oder auf den Jogginghosen-Boom zu reagieren [The Pricing Machine; Artificial Intelligence vs. Human Gut Feeling: How Data Helps Drive Higher Margins, Wean Off Bums, or Respond to the Sweatpants Boom]. *Textiwirtschaft*, 25 March 2021.

4. Source: Horváth (2022).

5. Source: Horváth (2022).

Chapter 8

1. Farouky, J. (2007). Why Prince's free CD ploy worked. *Time*, July 18. http://content.time.com/time/arts/article/0,8599,1644427,00.html

2. O'Reilly, Terry (2013). Loss leaders: How companies profit by losing money. *CBC Radio: Under the Influence*, April 20. https://www.cbc.ca/player/play/1616745539711 (accessed September 13, 2021).

3. https://fortune.com/company/alphabet/fortune500/

4. *The Guardian*, Many Ryanair flights could be free in a decade, says its chief. https://www.theguardian.com/business/2016/nov/22/ryanair-flights-free-michael-oleary-airports

5. Trony leaflet: Buy 3, pay 2, July 2021. https://www.mondovolantino.it/cataloghi/trony/volantini/1723

6. Liu, C.Z., Yoris, A.A. and Hoon, C.S. (2015). Effects of freemium strategy in the mobile app market: An empirical study of Google Play. *Journal of Management Information Systems* 31 (3): 326–354.

7. Microsoft website (2022). Earn rewards just for searching on Bing. https://www.microsoft.com/en-us/bing/defaults-rewards (accessed January 1, 2022).

8. Miles & More website (2022). https://www.miles-and-more.com/de/de/program/daily-benefits/milespay.html (accessed January 1, 2022).

9. *Outside* business journal (2008). Danish gym offers free membership — unless you don't show up. September 29. https://www.outsidebusinessjournal.com/brands/danish-gym-offers-free-membership-unless-you-dont-show-up/

10. Herstand, A. (2014). Should you pay to play? Here are the worst to best club deals in the world. *Digital Music News*, April 16. https://www.digitalmusicnews.com/2014/04/16/should-you-pay-to-play/

11. Free Conference Call.com website: https://www.freeconferencecall.com/international/de/en/?marketing_tag=FCCIN_PPC_GB_DE_EN_0044&gclid=CjwKCAjw7fuJBhBdEiwA2lLMYf35-4XY3zxKDPzMh5q 41D8nP742QnadVDPV5lYV6p_wFz4wAQ1em xoCpxUQAvD_BwE

12. Before the price model was changed.

13. Pietschmann, C. (2020). Microsoft for startups: Free Azure and other benefits. *Build5Nines*, January 14. https://build5nines.com/microsoft-for-startups-free-azure-and-other-benefits/

Chapter 9

1. Balch, Oliver (2015). Is sympathetic pricing anything more than a novelty? *The Guardian*, April 9.

2. Global trend briefing (2014). Sympathetic pricing. *trendwatching.com*, June. https://www.trendwatching.com/trends/sympathetic-pricing

3. Lies, Elaine (2014). Japanese men baldly go into new Tokyo restaurant, with pride. *Reuters*, May 9.

4. Walsh, Michael (2013). Petite Syrah cafe in Nice, France, bases coffee prices on patrons' manners. *New York Daily News*, December 11.

Chapter 10

1. Everlane, The annual *Choose What You Pay* event. https://www.everlane.com/choose-what-you-pay

2. *Business Insider* (2020). Everlane's big "Goodbye 2020" sale includes its ultra-comfy knit ballet flats, cashmere sweaters, and sneakers. https://www.businessinsider.com/everlane-sale-choose-what-you-pay?r=US&IR=T

3. *People* (2020). This Meghan Markle-loved brand is letting you choose what you pay today. https://people.com/style/shop-everlane-choose-what-you-pay-sale/

4. Baldwin, C. (2020). Everlane's "Choose What You Pay" sale is full of things you want right now. *Forbes*, May 13. https://www.forbes.com/sites/forbes-personal-shopper/2020/05/13/everlane-choose-what-you-pay-sale-is-full-of-perfect-summer-basics/?sh=3ddf8ace5cb0

5. Townsend, T. (2015). Everlane is letting customers choose prices through New Year's Eve. *Inc.com*, December 29. https://www.inc.com/tess-townsend/everlane-tells-customers-pay-what-you-want.html

6. https://www.breadpayments.com/blog/these-3-brands-succeed-by-letting-customers-pay-what-they-want/

7. https://www.breadpayments.com/blog/these-3-brands-succeed-by-letting-customers-pay-what-they-want/

8. Kim, Ju-Young, Natter, Martin and Spann, Martin (2009). Pay what you want: A new participative pricing mechanism. *Journal of Marketing* 73: 44–58. https://www.ecm.bwl.uni-muenchen.de/publikationen/pdf/pwyw_jm.pdf

9. https://londoncashmerecompany.com/pages/what-is-choose-what-you-pay

10. 7-Eleven: Name Your Own Price Slurpees on Saturday 11/7. Fortune (2013) Pay any price you want (even 1¢) for Slurpee at 7-Eleven on 11/7. *Fortune*, November 3. https://fortune.com/2015/11/03/free-slurpee-7-eleven-name-your-own-price/

11. Also known as customer-driven, interactive or subjective pricing.

12. Allwetterzoo Münster (2013). Fünf Mal so viele Besucher dank "Pay what you want". January 13. https://www.welt.de/regionales/

NOTES

duesseldorf/article113280670/Fuenf-Mal-so-viele-Besucher-dank-Pay-what-you-want.html

13. Abel, A. (2021). At this adorable Italian Hotel, the new idea is pay what you want and pay it forward. *Forbes*, May 24. https://www.forbes.com/sites/annabel/2021/03/24/at-this-adorable-italian-hotel-the-new-idea-is-pay-what-you-want-and-pay-it-forward/?sh=72b182566c14

14. Sosta sospesa – OmHom: https://www.forbes.com/sites/annabel/2021/03/24/at-this-adorable-italian-hotel-the-new-idea-is-pay-what-you-want-and-pay-it-forward/?sh=72b182566c14

15. Traveller.com (2009). New Singapore hotel offers "pay what you want" rates. https://www.traveller.com.au/new-singapore-hotel-offers-pay-what-you-want-rates-7zbt

16. Zahlt doch, was ihr wollt! Schmidtchen Theater, Reeperbahn: https://www.tivoli.de/service-kontakt/unsere-theater/schmidtchen/unbedingt/zahlt-doch-was-ihr-wollt/

17. Schauspielhaus Zürich (2022). Pay as much as you want. https://www.schauspielhaus.ch/en/1757/pay-as-much-as-you-want

18. Jazzy (2021). 13 pay-what-you-want restaurants around the world. *Road Affair*, December 8. https://www.roadaffair.com/pay-what-you-want-restaurants/

19. *Evening Standard* (2015). https://www.standard.co.uk/go/london/the-london-restaurants-that-are-letting-you-pay-what-you-want-a2324096.html

20. Salaky, K. (2020). Burger King is offering 'pay what you want' on whoppers today. *delish.com*, September 5. https://www.delish.com/food-news/a34163110/burger-king-whopper-pay-what-you-want/

21. https://support.humblebundle.com/hc/en-us/articles/204387088-Pay-What-You-Want-and-Contribution-Sliders

22. Feccomandi, A. (2020). The amazing success of the "pay what you want" model. *bibisco blog*, December 20. https://bibisco.com/blog/the-amazing-success-of-the-pay-what-you-want-model/

23. Red Dot Design Museum (2022). How much would you like to pay for your admission? https://www.red-dot-design-museum.org/essen/visit/admission/pay-what-you-want

24. Kalepa website: https://www.thekalepagroup.com/customer-experience-inspiration-sessions/

25. https://theedinburghreporter.co.uk/2021/08/leiths-pay-what-you-want-bookshop/

26. About Us – OpenBooks.com (accessed August 28, 2021).

27. Canadian Premier League (2021). Atlético Ottawa unveils "pay what you want" ticket offer for first-ever home match. *canpl.ca*, July 18. https://canpl.ca/article/atletico-ottawa-unveils-pay-what-you-want-ticket-offer-for-first-ever-home-match

28. https://www.nbcnews.com/business/consumer/new-activehours-app-lets-you-pick-pay-your-paycheck-n170791

29. https://michaelstipe.com

30. Bandcamp website: https://bandcamp.com/tag/name-your-price (accessed August 22, 2021).

31. https://www.garmentory.com/static/garmentory

32. Conlan, E. (2011). Is Gap "My Price" the new priceline for clothing? *SHEFinds*, May 12. https://www.shefinds.com/is-gap-my-price-the-new-priceline-for-clothing/

33. eBay (2022). Making a best offer. https://www.ebay.com/help/buying/buy-now/making-best-offer?id=4019

34. Booking Holdings Inc. *(BKNG)*: https://finance.yahoo.com/quote/BKNG?p=BKNG&.tsrc=fin-srch

35. https://www.statista.com/statistics/225455/booking-holdings-total-revenue/

36. 2021 https://www.priceline.com/static-pages/best-price-guarantee.html

Chapter 11

1. Was darf Kaffee kosten? *Süddeutsche Zeitung*, December 23, 2016.

2. Interview between Kai-Markus Müller and Danilo Zatta, February 2022.

3. Prelec, D. and Loewenstein, G. (1998) The red and the black: Mental accounting of savings and debt. *Marketing Science* 17 (1): 4–28.

4. https://faculty.washington.edu/jdb/345/345%20Articles/Iyengar%20 %26%20Lepper%20(2000).pdf

Chapter 12

1. HP website (2022). https://instantink.hpconnected.com/uk/en/l/v2 (accessed January 3, 2022).

2. McKinsey Digital (2015). Reborn in the Cloud. https://www.mckinsey.com/business-functions/mckinsey-digital/our-insights/ reborn-in-the-cloud

3. McKinsey Digital (2015). Reborn in the Cloud. https://www.mckinsey.com/business-functions/mckinsey-digital/our-insights/ reborn-in-the-cloud

4. McKinsey Digital (2015). Reborn in the Cloud. https://www.mckinsey.com/business-functions/mckinsey-digital/our-insights/ reborn-in-the-cloud

ABOUT THE AUTHOR

Danilo Zatta is one of the world's leading advisors and thought leader in the field of Pricing and TopLine Excellence. As a management consultant for more than 25 years, he advises and coaches many of the world's best-known organizations. He has led hundreds of projects both at national and global level for multinationals, small and medium-sized companies as well as investment funds in numerous industries, generating substantial profit increases. His advisory work typically focuses on programs of excellence in pricing and sales, revenue growth, corporate strategies, topline transformations, and redesign of business and revenue models.

He acted as CEO, Partner and Managing Director at some of the world's leading consulting firms, building up international subsidiaries, entire pricing and sales practices, and fostering growth. Dan has also written 20 books including *Revenue Management in Manufacturing* (Springer, 2016). He has also published hundreds of articles in different languages and regularly acts as keynote speaker at conferences, events, associations, and at leading universities. He also supports as personal topline coach several CEOs of leading companies.

ABOUT THE AUTHOR

He graduated with honors in economics and commerce from Luiss in Rome and University College Dublin in Ireland. He got an MBA from INSEAD in Fontainebleau, France and Singapore. Finally, he completed a PhD in revenue management and pricing at the Technical University of Munich in Germany.

Connect with Dan on LinkedIn. If you would like to talk to Dan about any advisory work or speaking engagements, please contact him via email at: danilo.zatta@alumni.insead.edu

INDEX

References to Notes will contain the letter 'n' following the Note number

INDEX

Printed in the USA
CPSIA information can be obtained
at www.ICGtesting.com
LVHW022023141023
760425LV00009B/210/J

9 781119 900573